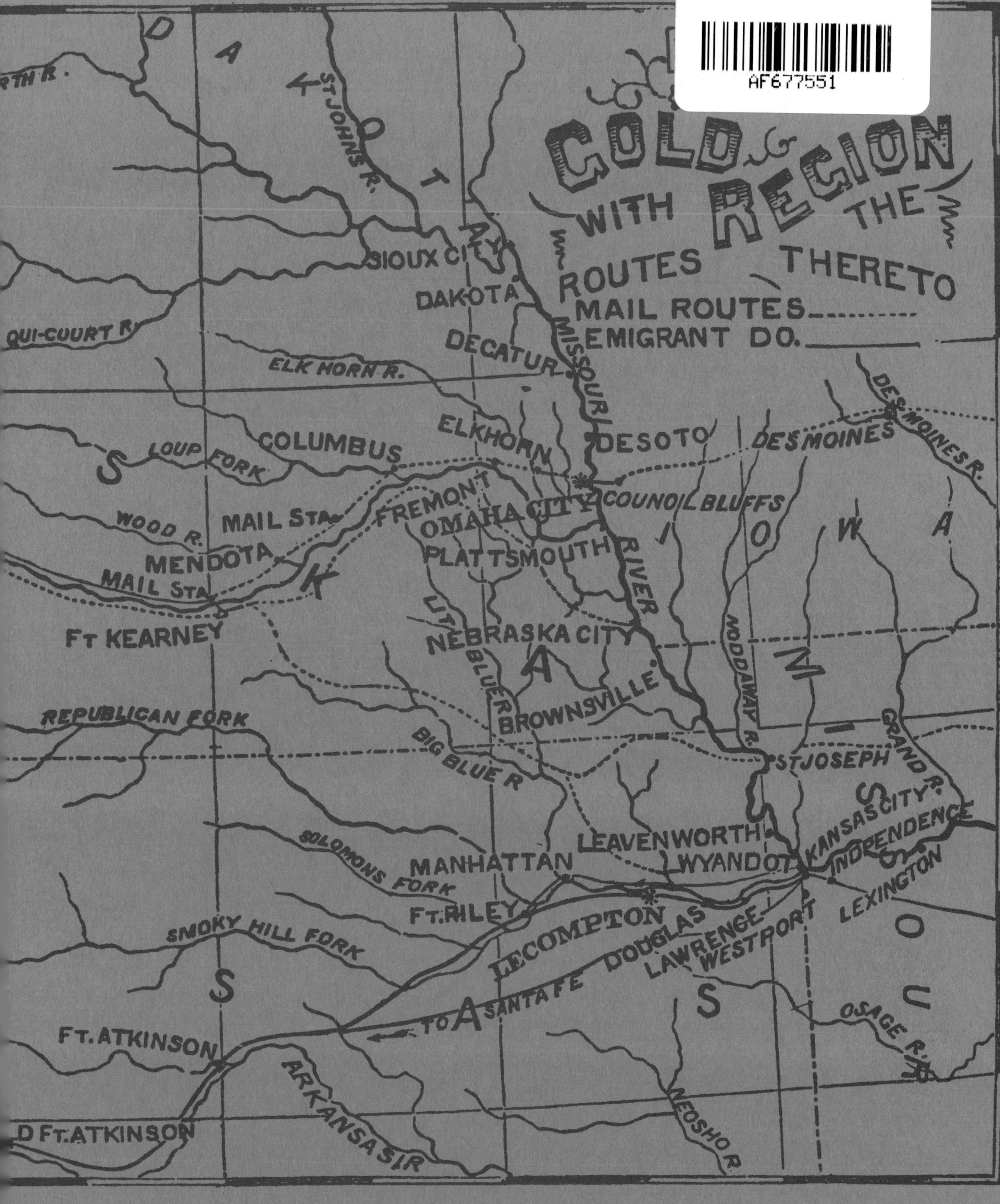
GOLD REGION
WITH THE
ROUTES THERETO
MAIL ROUTES
EMIGRANT DO.
DAKOTA
ST JOHNS R.
QUI-CUURT R.
SIOUX CITY
DAKOTA
DECATUR
MISSOURI RIVER
ELK HORN R.
ELKHORN
DESOTO
DES MOINES
DES MOINES R.
COLUMBUS
LOUP FORK
COUNCIL BLUFFS
FREMONT
OMAHA CITY
PLATTSMOUTH
WOOD R.
MAIL STA.
MENDOTA
MAIL STA.
Ft KEARNEY
NEBRASKA CITY
BROWNSVILLE
NODDAWAY R.
IOWA
MISSOURI
LIT. BLUE R.
BIG BLUE R
REPUBLICAN FORK
ST JOSEPH
GRAND R.
KANSAS CITY
INDEPENDENCE
LEAVENWORTH
WYANDOT
MANHATTAN
SOLOMONS FORK
FT. RILEY
LECOMPTON
DOUGLAS
LAWRENCE
WESTPORT
LEXINGTON
SMOKY HILL FORK
TO SANTA FE
KANSAS
OSAGE R.
FT. ATKINSON
ARKANSAS R
NEOSHO R.
D FT. ATKINSON

Colorado Mining

Colorado Mining

A Photographic History

Duane A. Smith

UNIVERSITY OF NEW MEXICO PRESS
Albuquerque

Library of Congress Cataloging in Publication Data

Smith, Duane A
Colorado mining.

Bibliography: p.
Includes index.
1. Mines and mineral resources—Colorado—History.
I. Title.
TN24.C6S6 338.2'09788 76-46583
ISBN 0-8263-0437-0

Manufactured in the United States of America. Library of Congress Catalog Card No. 76-46583. International Standard Book No. 0-8263-0437-0.
First Edition.

For
Robert Athearn
Lee Scamehorn
Carl Ubbelohde

Preface

After a climb to a mine site, it is pleasant to gaze upon man's relics and nature's beauty. Mountains stretch endlessly and even their barren tops show signs of mining—a hole here, a small dump there. The fast-growing aspen cannot completely mask the scars on the lower slopes, where man burrowed into the mountainside searching for the elusive minerals he dreamed would make him as "rich as Croesus." Where life once hummed, only the wind rustles the aspen and spins through the ruins of dreams. Such a scene inspires one to think about men, mining, and times past.

Mining, the foundation stone of Colorado, was the state's number one industry for more than a generation after 1859. Stories larger than life about mining abounded. Mining itself often loomed larger than life, with its possibilities of instant wealth, its lure of new opportunities and ever changing fortunes, and the expectation that the next blast or swing of the pick would uncover the "mother lode." Human nature took the possibilities and created legends. Eventually the era vanished; wishes and feeble attempts to reconstruct it were futile. In its place came industrialization and corporate operation: management and labor, wages and depersonalization, violence and strikes. This period generated its stories, too, but they were neither as romantic nor as popular.

Since 1859 men have been mining in Colorado (women have now joined them underground). The wealth they have brought out of the soil has enriched the state, the nation, and even foreign investors' pockets. There is more to the story than just the monetary returns from gold, silver, coal, or oil; mining spawned a life-style, dominated politics, encouraged agriculture and urban growth, fostered a transportation network, touched the lives of Coloradans in many and varied ways, and left behind a tourist bonanza.

No comprehensive history of Colorado mining has ever been written, and this is not meant to be the definitive examination. The objectives of this volume are to give insight into the industry's development and the lives of the people who provided the heart and soul that kept it going. It is also intended to encourage others to study various aspects of mining only touched upon here, but in need of further examination. Ideally the reader will finish with a better concept of what transpired, the effort involved, and the significance.

Developing simultaneously with Colorado mining was photography—from the glass plate to the handheld Kodak camera in a generation. Photographers came almost on the heels of the fifty-niners; the photographs they took show an unevenness in quality and, especially in the nineteenth century, the stiffness of posed scenes. Nevertheless, they can be appreciated for the story they tell. They supply a glimpse of the past that words cannot adequately describe. The viewer is transported to a sunny day in a far off land and is allowed to let his or her imagination roam.

I owe an unrepayable debt of thanks to many people who have shared their ideas and photographs with me. It is impossible to thank them all in print, but this does not lessen my appreciation of their kindness. The following were especially helpful: Allan Bird, Steve Frazee, Mrs. Homer Reid, Mrs. Marvin Gregory, Bluford Muir, Rob DeNier, Stanley Dempsey, Robert Richmond, H. C. Osborne, Bill Clannin, Gary Huber, Leland Prater, Robert Rinker, Dave Cole, Dayton Lummis, Ken Periman, Jack Eberl, Richard Gilbert, and Rich Yates. To the many staff people of the various libraries, agencies, museums, archives, and companies who searched for pictures and answered a host of questions, my sincere gratitude.

The University of Colorado Centennial Commission generously provided the grant which allowed travel and the purchase of the photographs. My special appreciation goes to Bud Arnold and Gerry Bean for their support of this project. Again, as they have in the past, several of my friends and colleagues provided assistance in various ways: Maxine Benson, Ned Blair, Morris Taylor, Roberta Schilling, Liston Leyendecker, Marguerite Norton, and Robert Delaney.

Remembering my school years brings to mind a number of outstanding teachers, from the kind ladies who taught the primary grades and those who awakened interest in history, down to the professors who directed early research and writing efforts. All contributed and I stand in debt to each; their concern, counsel, and help is probably more appreciated now than when it was offered. Three particularly enriched my educational experience through example and direction. With kindness and patience they instilled in me the tools of the profession, the discipline of scholarship, and a love for the past. To them this book is dedicated as only an inadequate token of what they have meant to me.

Contents

List of Illustrations

Prologue

Antiquity guards the secret of the first discovery of precious metals in what became Colorado. An Indian bending over a stream was probably the first; attracted by its glitter, he picked up a piece of shining rock, rolled it around in his hand, and moved on. It seems likely that a coal outcropping caught the attention of other inquisitive natives; not realizing its uses, they would also have discarded the black rock. No chronicler or reporter recorded their discoveries and these moments were lost to history.

Centuries later, men came from the south, lusting after that very gold the Indian had so casually examined. These men, Spaniards, knew the value of gold and were willing to sacrifice and fight for its possession. Coronado searched in vain, missing completely the mineralized regions as he traversed the Rio Grande Valley and the Great Plains. Other Spaniards followed him, and settlements soon took root where he had passed. From those settlements came the first European discoverers of Colorado gold and silver.

Who they were is unknown, where they mined can be guessed. In the southwestern part of the state they left behind names like Sierra de la Plata and Río de la Plata—mountain and river of silver. Wandering into the melancholy depths of a canyon, they christened its river Río de las Ánimas Perdidas, river of lost souls; it became a fitting epitaph for many who sought the minerals in the depths of the mountains. The Dominguez-Escalante party, traveling through the area in 1776, noted in their journal that mining had been undertaken earlier: "The opinion formed previously by some persons from the accounts of various Indians and of some citizens of this kingdom that they were silver mines, caused the mountain to be called Sierra de la Plata." Escalante was unsure of who, when, or what metal was involved, but he did refer to the Don Juan María de Rivera expedition of 1765. Mining, however, had been carried on before that time.

That the Spanish actually did prospect and mine in Colorado is supported by findings of early American miners. Abandoned shafts, cuts, tools, and other relics were found from the San Juan Mountains northeastward to the front range of the Rocky Mountains. Some of the reports may have been based only on the imagination; however, the total Colorado coverage and varying discovery dates authenticate the finds. In addition, the legends of lost mines and the maps which label this region as one having gold and mines make it obvious that, on a small scale, at least, mining activity was carried on during the eighteenth century.

With the opening of the nineteenth century, evidence to substantiate the existence of gold began to mount. In an 1804 report, Regis Loisel, a St. Louis resident, claimed there was gold on the Platte River, so abundant that "nuggets, scattered here and there" were found. His story was reinforced by James Purcell, whom Zebulon Pike encountered in Santa Fe, following Pike's seizure by Spanish officials in 1807. "He assured me that he had found gold on the head of the La Platte, and had carried some of the virgin mineral in his shot-pouch for months." If he actually had found gold, it was of little use to him because the Spanish would not let him leave New Mexico, and Purcell refused to show them where he had found it, since he believed the site to be in American territory.

In the years that followed, Spain lost Colorado to Mexico, which in turn lost it to the United States. Meanwhile, the mountain men and a few New Mexicans traveled through the land, leaving little to mark their passage. Josiah Gregg, for example, in his *Commerce of the Prairies* only briefly mentioned a very rich placer that had been discovered north of Taos above the thirty-seventh parallel. Stories of gold bullets and silver trinkets aside, the era of the mountain men did provide some intriguing questions. William Bent, of Bent's Fort fame, claimed the Indians knew of the existence and location of gold but kept the knowledge to themselves for fear of losing their hunting grounds. Bill Williams, a trapper who trekked through Colorado, returned home to Missouri in 1841, reportedly carrying "gold nuggets from the Rockies." Even earlier, in the 1830s, a small group of trappers and traders-turned-miners allegedly found gold near what was later Golden, Colorado.

The discovery of gold in California in 1848 launched a westward rush over the territory to the north and south of Colorado, but few people stopped in the region. One Oliver P. Goodwin claimed several years later to have toured the front range in 1849, finding a few "good prospects." These 49'ers were eager to move on to more promising streams and tarried only briefly. One member of a band of Cherokees, camped on what they called Ralston's Creek a few miles from the site of Denver, kept a diary and, on June 22, 1850, noted, "Gold found." However, it was not enough to detain the Indians for long, bound as they were for the Sacramento.

In the remaining years of the 1850s, before serious prospecting was undertaken, reports filtered east and appeared in various border town newspapers. The Lawrence, Kansas, *Herald of Freedom*, on May 26, 1855, reported "gold in great abundance" on the headwaters of the Arkansas River. D. C. Hail, traveling to California via the Colorado region, wrote a letter that was reprinted in the (Little Rock) *True Democrat*, April 13, 1858, telling of gold prospects on the upper waters of the Arkansas and Platte rivers. On the eve of the Pike's Peak discoveries, the Marysville *Daily California Express* (July 21, 1858) recounted "gold fever" raging in Kansas, the consequence of discoveries in the western part of the territory. The communication lag precludes this item's referring to the 1858 activities of the William Green Russell or Lawrence, Kansas parties, so it must have alluded to an earlier rumor.

The shining mountains, which had intrigued the Spanish, French, and English for centuries, cast their spell over Americans. Parties were on their way to investigate rumors; what they found in fact would chart the course of Colorado history. The question of why mining had not taken hold before this may be fairly asked.

Josiah Gregg provided part of the answer, when he discussed that rich placer north of Taos. The site, he noted, was among the snowy mountains and "ice bound over half the year." Isolation and a harsh environment discouraged would-be miners. So did the Indians, who, as Bent suggested, were not happy with the prospect of seeing their hunting grounds overrun. Ownership of the land proved uncertain before the Mexican War (1846–48) settled the issue. Spain did not welcome outsiders and had done everything within its power to discourage the encroachment of foreigners. Until the Louisiana Purchase, the United States was at a great distance, and even after, disagreement persisted over who owned what. Mexico's policy toward Americans vacillated and eventually led to its losing a vast amount of territory.

There were other deterrents to mining. Prior to the mid-1850s, conditions had not been right for a mining stampede. Discoveries had always been small and told after the fact; no excitement had been generated by firsthand or newspaper reports, accompanied by samples. Then the California bonanza created a climate of nearly uncritical faith in the treasure mountains of the West. With California as a precedent, it was easier to hurry off, knowing what had been found by those who had arrived in 1848–49.

By the late 1850s, isolation was being alleviated, territorial ownership had been decided, the Indians were less fearsome as army troops patrolled and forts were garrisoned, and improved communications engendered enthusiasm. A dash of gold excitement added the necessary spice, and the foundations for a rush were laid.

Gold and silver exerted an irrational hold on men's imaginations. For over three hundred years, legends, rumors, and a little ore had led the captivated toward Colorado. Undoubtedly, the most successful ones had been the Spaniards, who came up from the Rio Grande Valley. However, they left no known written records, preferring to take what they could mine in a season or two, leaving because of Indian pressure or before the king's agents appeared and they were compelled to give the king his royal share. They left behind legends of lost mines and buried treasure. In years to come others would chase the same elusive gold.

In a sense, they all were chasing a rainbow. More mines were veined with disappointment and shattered dreams than with gold or silver. Still, men persisted. Thoughts of all those minerals in the ground, just waiting to be retrieved and converted to wealth, drove them on. Mineral deposits haunted and teased, promised much, and yielded to only a precious few. Holding a piece of high-grade ore and seeing the glittering gold can infect anyone with a case of mining fever. A lifetime could be spent trying to cure the disease. Even the discovery of a paying mine seemed only to push the fever higher. Who can say what motivated these miners and prospectors? Perhaps it was greed—perhaps excitement and wanderlust; the reasons were nearly as many as the men who chased that rainbow. At any rate, somewhere beyond the next mountain, there was always that personal El Dorado.

The words of a folksong express as nearly as anything what was beyond that next mountain:

> Come, come, there's a wondrous land,
> for the hopeful heart, for the willing hand.

Colorado was a wondrous mining land. It still is, and this is its story.

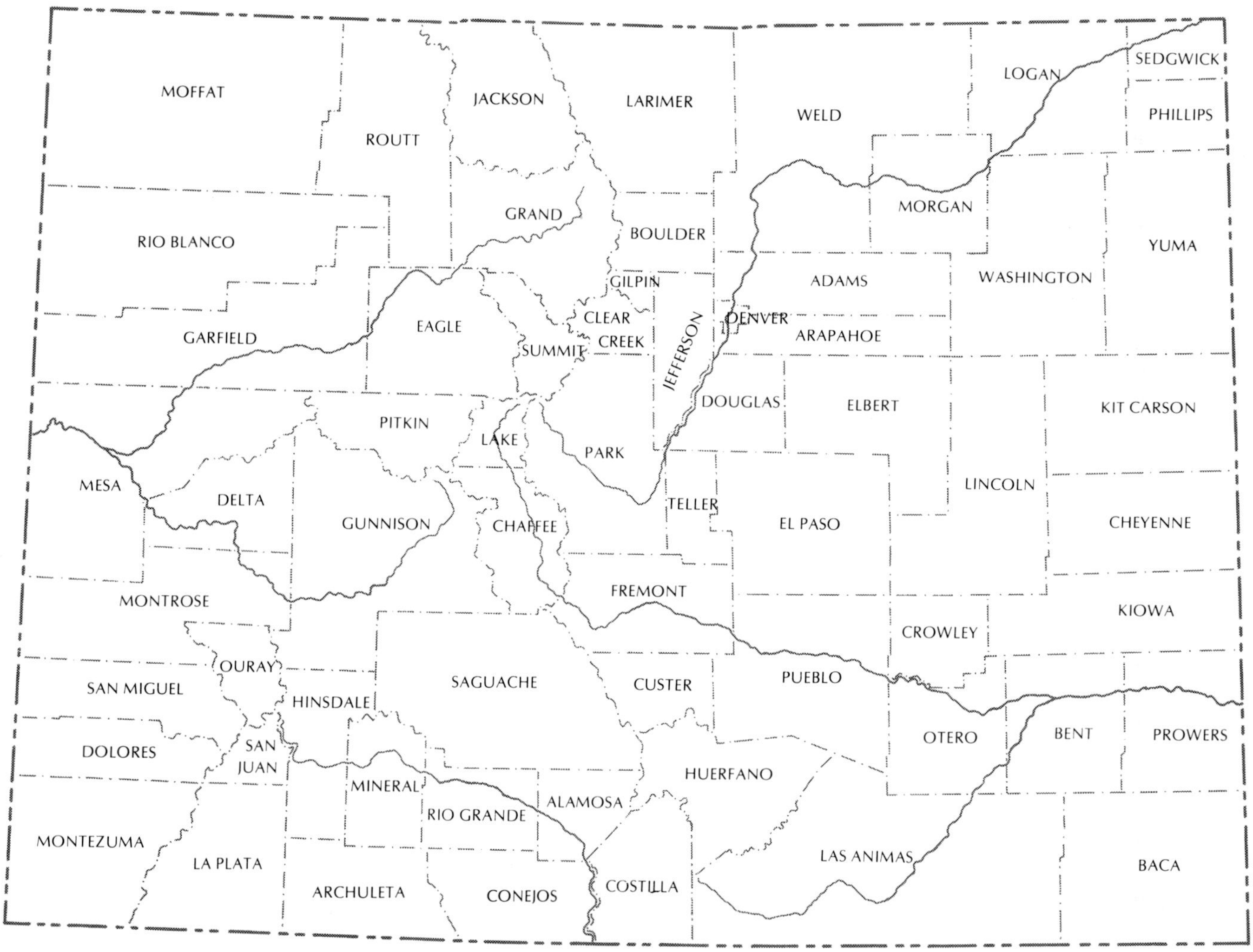

MAP 1 County Map of Colorado

1

Pike's Peak or Bust

"Pike's Peak or Bust." For a few months those words distracted Americans' attention from the cold war between the North and the South. As California had before, they exemplified what it was thought the West should be–a land of opportunities and abundant possibilities. They stirred men's imaginations and galvanized them into the second largest mining rush in United States history.

It was not to a strange, unknown land that these people were drawn. Pike's Peak country was becoming well traveled; fur trappers, explorers, hunters, army patrols, and even tourists had roamed its eastern front and penetrated the mountains beyond. A few parties of miners bound for California had panned some of its streams, but few stayed permanently. Settlement also had gradually crept closer, reaching into Kansas and Nebraska territories; these two divided most of the Pike's Peak country between them by government fiat. Its mineral potential had already been speculated upon and some of it had even been timidly tapped.

In the summer of 1858 two parties moved toward this land bent upon finding gold. One was the Russell party out of Georgia and Indian Territory. The hopes of this group were based upon the leadership of William Green Russell, an experienced Georgia and California miner, and the knowledge that Cherokees had actually found gold in the foothills of the Rockies while on their way to California. The party, numbering over a hundred persons, traveled along the Santa Fe Trail, then turned northwest at Bent's Fort, reaching the future site of Denver in May. Neither there at Cherry Creek, nor in nearby streams, did they find promising "color" in their pans. Discouragement set in and most of them turned dejectedly homeward. Russell, his two brothers, and a handful of friends stayed. In early July, they found a small pocket of gold at the mouth of Dry Creek. They soon exhausted its few hundred dollars' worth of gold, but buoyed by the excitement of discovery, they moved northward along the foothills.

Unknown to them, a second group was also on its way. Known as the Lawrence party, from the town in Kansas, they had been encouraged to come West by the stories of a Delaware Indian, Fall Leaf, who had passed through the Pike's Peak country in 1857 as an army guide. Fall Leaf had brought back some gold, a lure hard to resist. About thirty men made up this party, which began prospecting near Pike's Peak, then moved over into South Park, without any luck whatsoever. Moving down near the army post of Fort Garland in the San Luis Valley, this group heard of Russell's success and hurried north. At Cherry Creek the two parties met when Russell's men returned from their prospecting tour.

1. The rush to Pike's Peak was on. This fifty-niner drew a crowd of onlookers in St. Joseph, Missouri, as he prepared to travel west. Courtesy the Kansas State Historical Society, Topeka.

Although their hard work, time, and trouble produced little profit, gold rushes had been built on even less. Several people had visited the Russell party while it was panning out that one pocket; one of them was a trader, John Cantrell, who left with a small amount of placer gold. Cantrell hurried east, carrying, in addition to the sample, a vivid imagination, which inflated the importance of the discovery in the weeks between leaving the "mines" and reaching "civilization." On the basis of his story, the Kansas City *Journal of Commerce*, on August 26, ran the headline, "The New Eldorado!!! Gold in Kansas Territory!!"

The story was picked up by presses throughout the country and inflated with each telling. On September 20, the *New York Times*, crediting an interview from the St. Louis *Democrat*, reported that gold had been found at Cherry Creek "in all places" prospected, and predicted that a new gold fever was plainly at hand. A few expressed doubts; the St. Louis *Republican*, October 5, reported that the latest accounts from the mines (note how the term "mines" was now accepted) were unfavorable. Yet evidence, real and imagined, continued to accumulate and, finally, an editorial appearing on February 21, 1859, in the *Times* announced that there was "no longer room for doubt" about the discovery of "an important gold region."

Gold fever was clearly prevalent. Stories spread like a prairie fire. One dealt with a Dutchman in Council Bluffs, who was observed collecting a large supply of flour bags. Asked what he planned to do with so many, he replied that he intended to "fill them with gold at Pike's Peak." Replying to scoffers, this undaunted 59'er emphatically stated he would do it, "if I have to stay there till fall." Apocryphal or not, there were indeed many Dutchmen with gold dust dancing in their eyes.

Papers were printing advice on how to reach the gold fields, what to take, the necessary mining skills, and other helpful hints, including this gem on how to treat Indians: "Treat them courteously, but not too familiarly. Should one 'deliberately' insult you, knock him down, but be chary in the use of fire arms." Guidebooks foisted all kinds of information—and misinformation—on the unsuspecting public, and river towns vied to promote themselves as the gateway to the mines. There was profit to be made in a gold rush and merchants did not let opportunity elude them.

It was not simply the lure of gold that enticed men. The Midwest was still suffering from the depression which had begun in 1857 and, with opportunities slim at home, Pike's Peak prospects looked all the better. Then there were always those who suffered from wanderlust and seized any rumor of this nature, leaving everything to gamble on something better out West. For whatever reason, most prepared to leave in 1859; a few jumped the gun and set out the previous fall—these were the 58'ers.

One Samuel Curtis succumbed to the gold fever; according to a friend, he was "almost carried away. . . . He will not lose much at any rate as he is making nothing here." Consequently, Sam journeyed west in the fall, becoming a 58'er. Like others, he rushed to try his hand at mining: "it [is] very hard work and after I had washed one pan full I concluded it was here, but that I was not adapted to digging it." He then turned to merchandising for his path to fortune. In 1859, from Council Bluffs, Iowa, his father encouraged his son's sense of history:

> You are engaged in a great work, that of pioneering and carving out a young empire. . . .Your position has turned the eyes of thousands upon you, and you should therefore be constantly nominated with a desire to act well the part which you have attempted to perform.

Sam stayed and never lost his mining interest; within a year he was writing of moving into the mountains in the spring and "going to mining."

Meanwhile, the members of the Russell and Lawrence parties were camped on Cherry Creek, little suspecting what a rumor-created bonanza they were sitting on. With cold weather fast approaching, they were setting up winter quarters when the first of the 58'ers arrived. Looking around them, they saw neither a golden bonanza nor much hope for one. Disillusionment might have put an end to the furor then and there but for the fact that winter storms curtailed overland travel and prevented the news from getting out. Unable to make their fortunes by mining, the people turned to town planning and promoting. When the 59'ers arrived, they would find plenty of town lots for sale, if nothing else. The only other comfort to these men during those long, cold winter months was the knowledge that, come spring, they would have the best jump on prospecting.

During the winter, a few hardy prospectors ventured into the mountains in an attempt to find the place from which gold washed down into the streams. It took both courage and perseverance for those men

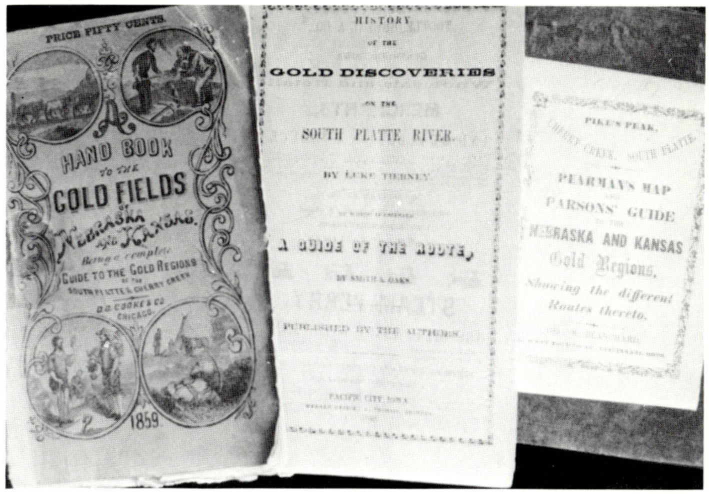

2. Using guidebooks such as these, the fifty-niners set off to seek their fortunes, trusting that the authors knew what they were describing. From the author's private collection.

to journey there; their dedication eventually paid off and gave credence to rumor. An experienced miner, George Jackson, discovered gold near future Idaho Springs in January 1859; he kept his secret well until spring. The same month a group of men found gold above Boulder at what they would call Gold Hill. Finally, in April, John Gregory hit the biggest discovery yet, between soon-to-be-established Central City and Black Hawk. These three discoveries gave a firm foundation for the gold rush which, in fact, was already underway.

Out came the 59'ers—hundreds, then thousands—some starting as early as February, long before the grass or weather was such as to encourage travel. Exactly how many came will never be known. Perhaps the number was as high as 100,000, of whom half might have persevered and reached Denver (as the settlement on Cherry Creek became generally known). Probably half of those decamped after a few discouraging weeks of hard and cold work prospecting the streams. The cost of living was so high that few could maintain themselves without finding either gold or a well-paying job. The remaining 25,000, more or less, became the basis for permanent settlement and the mining industry.

They came over all types of trails, in overloaded wagons, with pitifully poor maps showing nonexistent water holes. William Byers, who would have a long career in Colorado, wrote in his guidebook, "It is not the object of this work to persuade you to go." Many of the guidebooks were not that honest. They portrayed an easy trip, almost like a country ride. Most of the 59'ers rode in wagons or on horseback, a few walked, some pushed handcarts, and one dreamer lured a party aboard his wind wagon. This real prairie schooner, complete with sails, sank forever when it

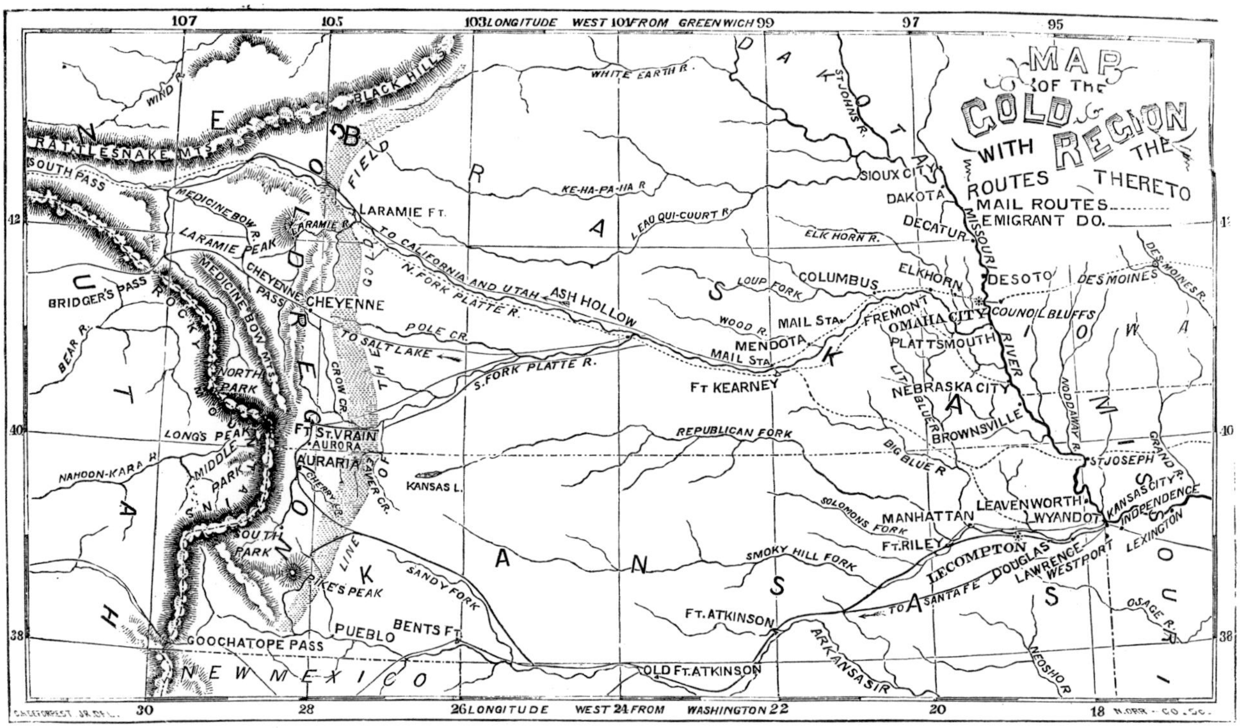

3. With maps like this, fifty-niners were sometimes led only to tragic ends. However, since this one recommended the established Oregon Trail up the Platte River, all probably went well for those using it. From the author's private collection.

encountered a gully not far from port. They came —gullible, excited, in a hurry.

Tales of hardships, including cannibalism, soon drifted back east. J. A. Wilkinson, coming West, was one of those who encountered men dragging back on foot without blankets or food and begging for something to eat. As far as he and his party were concerned, the Pike's Peak gold mines were "humbug," and they went on to California. One who returned home ("Go-Backers" and "Pilgrims," they were called) wrote that nearly all who ventured out were ruined. It was, he wrote, a "speculation" gotten up by businessmen in "the Territorys of Kansas and Nebrasky." Stories such as these, plus the fear of loss of population, angered the Chicago *Press and Tribune* enough to write on March 28, "We believe there is more gold to be dug out of every Illinois farm than the owners will ever produce by quitting the home diggings for those on the headwaters of the Arkansas and Platte."

They were right and they were wrong. Too much was expected from too little work. Gold was there and could be found. Byers, now editor and owner of the *Rocky Mountain News,* wisely commented in his first issue: "If we credited individual stories we could relate marvels equal to the Arabian Nights for wealth. . . . But from the mass of marvel we can glean this truth. Gold, in scale, exists, in sufficient quantity to reward the working miner. . . ." In another article he stated hopefully: "We know enough, however, to believe that a large population will settle here at once, and prosper, and we should believe this will be a reading and intelligent population."

Attention during the spring of 1859 centered on the area around Jackson's and Gregory's discoveries; men and a very few women hurried there. Horace Greeley, on a cross-country trip, stopped in Denver to cover the gold-rush excitement, joined them at Gregory's diggings in May, and left this impression:

> As yet, the entire population of the valley —which cannot number less than four thousand, including five white women and seven squaws living with white men—sleep in tents, or under booths of pine boughs, cooking and eating in the open air. I doubt that there is, as yet, a table or chair in these diggings, eating being done around a cloth spread on the ground. . . .The food, like that of the Plains, is restricted to a few staples —pork, hot bread, beans and coffee. . . .

From this came the Colorado mining industry.

The farmers, shopkeepers, teachers, and what have you—very few had any experience in mining—found the work of searching for gold taxing and not always remunerative. For every claim on which gold was found, there were others that yielded only disappointment. The 59'ers were looking for free gold, gold that could be recovered by the panning methods used in

4. A sluice box might exceed a hundred feet and needed a crew of men to operate, as shown by this 1860s scene in Gregory District, Gilpin County. More gold could be recovered this way than by panning; hence it was quickly adopted for use in the placer districts. Courtesy the Denver Public Library, Western History Department.

5. Central City as it probably appeared in the late summer of 1859. Log cabin architecture predominated, but false-fronted buildings were already appearing. Central was the hub of mining at this time. Courtesy the Denver Public Library, Western History Department.

California. A man needed only a pan, pick, shovel, and a strong back and constitution to be off on the search. With practice, the skill of swirling the gravel around the pan so that the sand and waste rock separated from the gold could be quickly learned. Keeping the pan tipped forced the heavier gold to the bottom, and it could then be taken out with tweezers or by mixing it with mercury. Mercury has an affinity for gold, and it easily separates the latter (in its free element) from surrounding materials. This was the basis of the amalgamation process that was long popular in Colorado. By heating the mercury—gold amalgam—the miner could drive off the mercury, leaving only the desired gold.

These placer operations along the streams and in the gulches were termed "poor man's diggings," because, as mining reporter and historian Frank Fossett later wrote, "little or no money is required to test their value or put them into producing condition. Every man knew what his claim was yielding when night came." Another writer put it this way, "The expressions of satisfaction or disappointment in those early Colorado mining times, when the sluice boxes were cleaned, would challenge the greed of the miner and the disgust of the spendthrift." Then and for the next few years the "poor man" was offered his best opportunity.

Panning was not only hard, but it was slow as well,

6. To operate a placer claim took water, which soon came to be in short supply. A Mountain City resident wrote in 1859 that good claims were available but they were dry claims "that cannot be worked without water." The Consolidated Ditch Company provided a remedy—for a price. Courtesy the State Historical Society of Colorado, Denver.

and again the California experience proved valuable. Soon rockers, looking much like a baby's cradle, were constructed. With these the washing process could be speeded up and more gold recovered. Longer and larger toms and sluice boxes soon followed, all utilizing the principle that the weight of the gold would cause it to settle as it washed down the troughs and allow it to be caught by cleats. In each case, however, a final panning and use of mercury were needed to free the gold from the remaining residue. The coarser gravel and sand were simply washed away in ever greater amounts.

With no organized property rights or governmental law to rely on, the miners turned to expedients tried elsewhere in American mining—miners' meetings, miners' laws, and mining districts. These extralegal controls gave some semblance of order to what would otherwise have been chaos. Initially the call went out for a miners' meeting, at which, in the fashion of frontier democracy, a chairman was elected and opinions voiced about size of claims, water rights, and other problems. Then perhaps a committee or two, or just individuals, drew up a set of laws to govern the district. The district was defined by prominent geographic features and usually was surrounded on all sides by similarly defined areas. The first mass meeting in the Gregory diggings was held June 8, 1859. None other than that famous visitor Horace Greeley admonished the reported two to three thousand assembled miners to avoid the temptations of drinking and gambling. After three rousing cheers and a couple of other speeches, the group got down to serious business. This included defining district boundaries, limiting the number and size of claims a miner could hold, specifying the amount of work necessary to hold a claim, and designating procedures for settling disputes. Then a district secretary was elected to record claims and call meetings. Miners from the Gregory district took their governmental experiences with them as they fanned out searching for gold. Other districts were formed; some elected more officers—constables, presidents—and formulated more laws as needed. Generally, all the miners wanted was something to guide their efforts and protect their claims. The simplest and cheapest method was to them the best.

Complaints and arguments over water rights nearly replaced boundary disputes. Water was required for all of these placer mining methods, and the streams quickly resembled liquid mud pies rather than sparkling mountain creeks. Soon, only those miners near the headwaters had clean water. Nor was there enough water to go around after the summer sun had melted the snows and the rains did not come. The miners quickly hit upon the idea of digging ditches and bringing water in from outside sources, rather than hauling the "dirt" to the water. The first ditch was built near Denver; a more famous one was built to serve the Gregory diggings. The latter, the Consolidated Ditch, when finished in 1860, was eleven miles long. Digging ditches in the mountains was hard and costly work; the 59'ers were finding out about mining in short order.

Another fact of mining life was the need for capital and manpower to work the claims. It was logical that if several men collaborated, they would have more of both. Companies started operating several claims, building larger sluices, and constructing short ditches to run the water through for greater pressure. A few built boom dams, collected water behind the dam, then let it go to loosen great quantities of "dirt." Efficiency was not inherent in this method. Then, in 1860, primitive hydraulic mining, which shot water through hoses and nozzles to wash the gravel down from stream banks, was introduced. It created a terrible mess but was faster and could be controlled better.

Men wise in mining knew that the placer gold had to be washed out of the mountainsides, and they set about following "float" to its source—the "mother lode." When they found an outcropping of likely looking quartz, they removed the ore with a pick and shovel, then panned to see what they had discovered. They were on the threshold of hardrock or quartz mining; if the gold was there, it would have to be dug out. They started like gophers and soon realized the need for blasting and timbering. As the shafts went down, whims and windlasses were necessary to bring the ore to the surface. If the miner were lucky, he could follow his vein via a tunnel or, more correctly, an adit. Mining became more costly and complicated every day.

When the ore was brought to the surface, it required crushing before it could be worked by the placer methods. First, primitive stone mortars, called *arrastras* by the Spaniards, were used whereby the ore was ground between two large rocks. Slow and cumbersome, generally horsepowered, arrastras did not provide the best solution. Stamp mills, already used in California, were the answer, and by the late fall several were in operation, and others were on the way. Here heavy weights, run by water or steam power, hammered the ore into a pulverized form that could be handled by the methods already discussed.

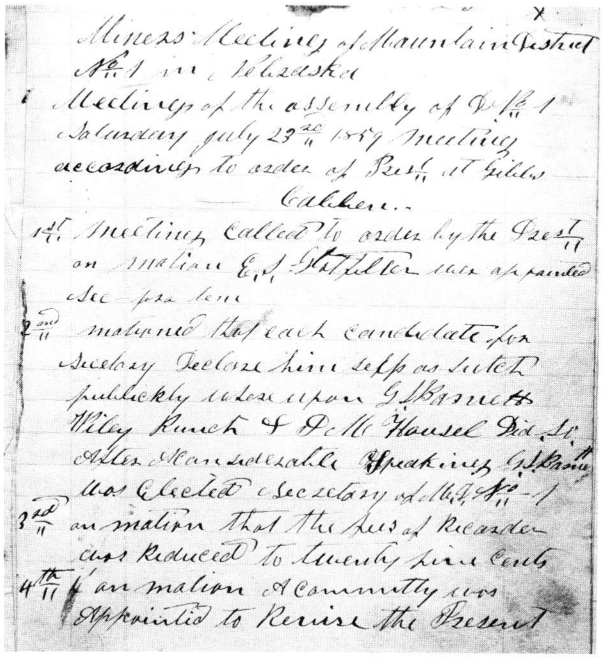
Miners Meeting of Mountain District No 1 in Nebraska
Meeting of the assembly of Dist No 1 Saturday July 23rd 1859 meeting according to order of Prest [illegible] Calkens
1st Meeting Called to order by the Prest on motion E. J. [illegible] was appointed Sec pro tem
2nd motioned that each candidate for Secretary Declare him self as such publickly whereupon G. J. Barnetts Wiley Runch & D. M. Hausel Did So. After considerable speaking G. J. Barnett was Elected Secretary of M. D. No 1
3rd on motion that the fees of Recorder was Reduced to twenty five Cents
4th on motion A Committy was Appointed to Revise the Present

7. The miners, without recourse to regular law or governmental agencies, took matters into their own hands and organized districts. This Boulder County meeting elected a district secretary and reduced the recorder's fees for filing claims. Courtesy Western Historical Collections, University of Colorado, Boulder.

As more and more people crowded into Gregory's and Jackson's diggings, the opportunity for profitable individual enterprise declined. Exploring parties began looking for new discoveries. They found them nearby in Russell Gulch and over in South Park, fifty miles away as the crow flew but somewhat farther by foot or mule. Prospectors also crossed the divide into the Blue River drainage and fanned out from the Boulder-Gold Hill diggings. Always rumors of more and greater finds swept ahead of them. Gold, silver, possibly precious stones—anything seemed possible in that wonderful year of 1859.

Not everyone turned to mining. An unidentified writer, quoted by Ovando Hollister in his 1867 *The Mines of Colorado*, had this to say. "Every one seems to think there is an easier way to make money than by digging, so that all other enterprises than mining are being overcrowded. I think they will find out to their sorrow that the gold has to be dug out of the ground before it can get into their pockets." The 59'ers were by no means self-sufficient; they needed to buy almost everything to underwrite their scrambling search for gold.

Hardly a mining district opened that did not spawn at least one camp, sometimes several. Crowded into the Gregory diggings were Central City, Black Hawk, Nevadaville, and Mountain City, with others not far away. Supply towns sprang up almost immediately at the gateways to the mountains; Boulder and Denver were prime examples.

Mostly jerry-built and often crowded together, the camps were not much to look at. Logs, tents, even dugouts and wagons, were pressed into service in the rush to open a business. A former Kansas City merchant, P. P. Van Trees, wrote his impression of Central City in August 1859: "I have not, as yet seen all of the city, it is scattered along the gulch, as far as I have seen, about two miles, quite compactly built; some very good log houses. We have some eight stores, many groceries, numerous bakeries, any amount of eating houses, one Masonic Hall and *nary* [*a*] *church.*"

Not far behind the merchant came the saloon keeper, perhaps a gambler, and others who would fill the miners' many and various needs. Some of the 59'ers themselves, disgusted with the low returns from their hard work, turned to occupations and professions they had pursued back in the States. All trades were in demand, from carpentry to farming to wood cutting. In some cases it was only a matter of weeks before the virgin forest and mountain gave way to settlement—settlement that was more than just a rough camp.

Hal Riley, writing to his father, expressed what must have been the common attitude that brought this all about: "A great portion of the emigrants came here with the expectation of picking up the gold like stones;

of course, they were disappointed. . . .We think we have got a good thing there [his claim]. I haven't made but little as yet. . . ." It was the Hal Rileys, the 59'ers who stayed in spite of all the work and obstacles, who forged a mining settlement out of mere hopes and dreams.

The mining frontier ebbed and flowed throughout the summer and fall of 1859. As winter closed in, the people drew back to the mountain settlements and even farther, to the towns nestled along the foothills. The year had not been as bountiful as the optimists had prophesied in 1858–59, nor had it been as dismal as the scoffers had forecast during the dreary days of the previous spring. The Pike's Peak rush engendered settlement and mining. While settlers and miners remained vulnerable, dependent as they were on long overland freight lines back to the "States," the cost of living, which had been high, now gave promise of stabilizing.

Most of the activity throughout the spring and summer had centered around placer mining. Neither as rich nor as extensive as those of California, the deposits that were found nevertheless put money in the bank for those fortunate enough to hit the rich claims. Already mining was changing, though, becoming more costly, difficult, and technical. From the simple pan and shovel, the industry had progressed to hardrock techniques and simple milling. The extent of the mineral deposits was as yet unknown; reports seemed to indicate that they extended far greater distances than originally estimated. Gold had been the cornerstone and was the only paying mineral at the moment, rumors and expectations notwithstanding.

The last issue of the *Rocky Mountain News* for the year, dated December 28, 1859, contained optimistic letters that talked about mills, tunnels, improved rockers, and mining which was "rapidly progressing." One writer complained that Denver merchants were attempting to depress the value of "gulch gold," thereby cheating the miners—a common complaint. Jealousy of Denver came early, as did a lingering suspicion that the miner was at the mercy of the nonmining population. Casting aside such concerns, a correspondent analyzed what he thought had happened in the past mining season and what needed to be done to prepare for the next: "Had those who came last summer had the perseverance, and sense enough to have pursued a like course [careful prospecting and hardrock mining], there would have been more leads opened, more gold extracted, and less stampeding and cries of humbug." In truth, there would have been, but it was the 1860s that would offer that opportunity.

2

Growing Pains

The rush of 1859, after a winter's pause, spilled over into 1860. As was the case in California in the 1850s, immigrants kept coming. Samuel Mallory, coming over the Platte route that spring, reported, "I think we see as many as 50 wagons per day." It is interesting that these later arrivals have never realized the same attention as their brethren of 1849 or 1859. Newspapers turned their attention to fresh, new subjects, and interest in the Pike's Peak rush gradually waned.

The 1860 rushers came to a much better defined area, of known wealth, that had been settled, amazingly enough, in only a year's time. They could traverse trails packed down by a year's travel to districts already proven, and to rough-edged settlements clinging to hillsides and gulch bottoms. But the welcome mat was not always out in the famous Gregory diggings. Ovando Hollister, early historian of this era, looking backward from the mid-1860s, wrote that it did not seem strange that the "old settlers" were somewhat cold with the immigrants. "They felt that they had earned what they had got, and that there was chance enough for others to do likewise. Surely, they said, all these strangers cannot expect employment here on our ground; let them branch out and find mines for themselves, or if not, go back."

The newcomers had several choices. They could work for someone else in one of the established districts, or purchase one of the claims already crowding the market. The latter was risky, since most newcomers possessed no knowledge for judging or working their investment. If they found mining not to their liking, they could always turn to one of the other occupations incidental to the growth of a mining region. Yet it was not to work for someone else, or for a humdrum job that promised steady employment and a living wage, that they had come West. These pioneers, like the 59'ers, wanted instant wealth. The question was where to find it. New districts just being opened, or prospecting to find one's own El Dorado, were two possibilities. Prospecting turned out to be something less than fun. As Hollister recalled, "[It] is a discouraging business except to the prospector by nature, who must have the faith of a martyr."

The best find in 1860 took place in the upper Arkansas Valley at a place called California Gulch, so named because the amount of gold found there reportedly resembled the discoveries of 1848–49 in the Mother Lode country. Whether Abe Lee, one of the discovery party, actually exclaimed that he had all of California in his pan when he found color there, matters little; he had found gold and stirred fresh excitement. Located near what would be Leadville, California Gulch proved to be the biggest mineral discovery in Colorado in 1860. News of it spread to nearby miners, then outward in an ever widening circle to eager listeners. On they came to the "richest gold deposit ever seen." Some older districts were entirely deserted; all suffered some loss of population. A mining district was organized and mining laws established, in the now time-honored fashion. Absentee ownership was forbidden, a practical solution to a problem which had plagued Gregory diggings. Working miners, not speculators, would have the best chance here.

In June a correspondent to the *Rocky Mountain News* wrote that the gulch, nearly ten miles long, was being prospected and mined, with claims selling for as much as $8,000 each. (So ephemeral, however, was the district's hold on the rushers that by summer people were already leaving California Gulch in search of "richer" deposits.) Pans gave way to sluices on the richer claims. Men organized themselves into companies in order to work more claims, secure more capital, and employ more manpower to make more money.

Throughout most of the summer miners kept coming; estimates ranged as high as 20,000; less than half that number would be more reasonable. Most found only disappointment, not gold, in their pans and left discouraged, either to try their luck somewhere else or to return to the States. Along the gulch a settlement grew, not planned but practical; its name was Oro City. It provided the miners with general merchants, blacksmiths, saloons, a bowling alley, and other required services. Webster Anthony, who arrived in Oro City in July to set up a general store, wrote in his journal:

> The streets appear as though every one built his cabin in its own place without regard to survey and as a consequence they are very crooked . . . . Have heard and read much about the rapid growth and population of these "fast cities" but to appreciate them one must see for themselves.

As fall turned to winter, many of those who had rushed there so eagerly a season before, departed. Oro City and California Gulch had their day. To be sure, the excitement carried into 1861, but with the placer deposits rapidly declining, attention was drawn elsewhere. Probably more than a million dollars' worth of ore was taken out in the first couple of years; no accurate figures are available. A few of the rich claims paid $1,000, or better, per day. Such was the nature of a placer district, though, that by the mid-1860s California Gulch was nearly deserted and forgotten. Only the man-made relics littered about the site served as reminders of what it had been.

The success of the Gregory diggings and of California Gulch kept the prospectors and miners on the move throughout the early 1860s. Few doubted that the next pan would uncover the bonanza, so off they went in the spring, not slowing down until cold and snow stopped practical operations. The enthusiasm of the times was such that they thought they had also discovered silver, diamonds, and opals, and most were confident that other precious stones and minerals would be found in this mineral treasureland. They went everywhere at the mere wisp of a rumor and too often came back poorer and not a whit wiser. Such was the nature of mining fever.

One of the most improbable rushes occurred in 1861 in a stampede to the San Juan Mountains. Located in the far southwestern corner of the territory, this area was at that time just about the most isolated, inaccessible, and even inhospitable spot in Colorado. That did not stop these Coloradans. In 1860 Charles Baker and his party had come into the San Juans as far as Baker's Park (the future Silverton area). Baker had found enough placer gold to excite him, and that winter his letters and enthusiasm sparked interest in the mountain towns and Denver. In the spring, prospector/miners set out from Oro City, Denver, and elsewhere, swinging southward into New Mexico and then northward up the valley of the San Juan River. Then they turned westward to the Animas River Valley, where Baker and his friends had thoughtfully built a toll road and laid out a town site. Over the mountains they trudged to Baker's Park to find plenty of hardships and hard work, but not much gold.

Provisions were short, costs high, and the weather cool, even in summer. Returns ran half a cent or only pennies per pan, maybe forty to fifty cents per day, not enough to live on. To make matters worse, the Utes were upset by this invasion of their land. A few disgruntled San Juaners wanted to hang Baker, who made the mistake of staying in his diggings; however, cooler heads prevailed and after a short season of prospecting, Baker's Park was abandoned. One distressed San Juaner returned to his old diggings at California Gulch with only moccasins, a blanket, and the clothes on his back. His experience caused him to resume mining there with renewed vigor and determination.

Nearly a decade would pass before people returned to the San Juans. Fortunately for Colorado, most mining experiences did not turn out this disastrously. At the same time that Baker and his followers went scurrying off, others became equally excited over the Buckskin Joe diggings in South Park. Although the area had been prospected as early as 1859, a year passed before lasting interest was generated in the fall of 1860. The rush came the following spring, bringing fewer people than California Gulch had attracted, but several times the reported five hundred who had gone to the San Juans.

8. An imaginative arrangement of sluice boxes and water flumes dominates this Gilpin County scene. It is doubtful that the man in the high hat and suit did more than pose. A sluice in rich gravel could clear between $400 and $500 per day; the average yield was much less. Courtesy the State Historical Society of Colorado, Denver.

Started as a placer district, Buckskin Joe was fortunate also to have quartz mines, which prolonged its existence until 1868. It was here that one of those enduring, perchance endearing, fables of Colorado mining history got its start. The legend of Silver Heels, the beautiful prostitute with a heart of gold who nursed miners during a smallpox epidemic, at the cost of catching it herself and ruining her "fabled" beauty, refuses to die. Romanticism clouded reality early on the Colorado mining frontier. What happened at Buckskin Joe paralleled what happened in the rest of the territory. When the placers gave out, the miners turned to quartz mining. They soon found out that it took both skill and money. The most easily worked ore near the surface was quickly mined and milled. Then came worries over saving the gold and the mounting costs of deeper mining. Financially unable to underwrite development, Buckskin promoters tried to sell claims to eastern investors. The few sales that were transacted only temporarily revived the district; money was spent faster than returns came in, dampening investors' enthusiasm posthaste.

Yale professor William Brewer, on a tour of the area in August 1869, happened upon Buckskin Joe and spent a few minutes wandering through the old town, peeping into windows.

> Such places have a sort of fascination for me—the old signboards in the streets, the roofless houses, the grass growing in the old hearthstones and flowers nestling in the nooks of mudded walls or broken chimneys, the multitudes of empty fruit cans, sardine boxes, etc. lying in the streets, bits of old saddles, rusty prospecting pans, old shovels, the stamps from mills—in fact, all the varied implements of a city, rusting and rotting, neglected.

A mining camp's life was as short as that of the mines around it. Buckskin Joe's story would be repeated many times in Colorado before the mining frontier period passed.

California Gulch, Buckskin Joe, and the San Juans were only three of many places affected during the gold rush era which had opened in 1859 and tapered off noticeably by 1862. By then much of the eastern and central mountains had been walked over, panned, and dug into with diminishing results. Coloradans never became blasé about new discoveries—they just learned from experience. The returns, especially when compared to those rich early days of 1859, did not justify rushing off here and there. Colorado's placer period proved much shorter than California's and its deposits nowhere matched the latter state's total production.

The questions raised by Buckskin Joe's quartz mining needed to be resolved if mining were to gain a sounder, more permanent, footing in the territory. Further, if any of the outlying districts hoped to prosper, transportation had to be improved. Attention focused on Gilpin County and its Gregory diggings; here at the most famous and thoroughly developed district existed the best chance to overcome these problems.

While gold fever was carrying people to all corners of the mountains, Gregory diggings and its camps settled down and population concentrated in Black Hawk, Central City, and Nevadaville. The *Tri-Weekly*

9. This is probably the Bobtail Gold Mining Company's property on the Bobtail lode, one of the famous early veins of Gilpin County. Early development featured uneconomical small companies working claims, some of which proved barren. Note the two whims in the center for hoisting the ore buckets and how the hillsides are already (c. 1864–65) denuded of trees. Courtesy the Denver Public Library, Western History Department.

Miner's Register, July-August 1863, reported that Central City had a dancing academy, a ladies' "Ice Cream Saloon," and two resident dentists. Black Hawk even had a hackney. Civilization had arrived. Local theatergoers could attend the Montana Theater's "Grand Spectacle—The Sea of Ice."

> The terribly grand scene of the breaking up of the sea of ice, the bursting of the storm, the frantic mother's prayer for her infant, the aurora borealis; all contribute to hold the audience spellbound until denouement.

The placer deposits were disappearing rapidly. Miners now burrowed into the hills, bringing out ore that, at first, was easily worked in nearby stamp mills; even arrastras were being pressed into service. After this ore zone was passed, the miners hit what they termed refractory ore, meaning that it resisted amalgamation. At a depth from sixty to a hundred feet, every miner found himself with ore that assayed well, only to find that the mills were unable to save enough to produce a profit. Whatever silver and copper might have been present flowed onto the dump or into the stream with the tailings. Some mills saved only fifteen percent of assayed value. Miners blamed the millmen, millmen blamed the ore, and nobody was happy. The inexperienced miners and millmen were mystified as to just what was going wrong. Commented one writer: "Somehow the mills as a general thing do not save the gold; why, it is hard to tell."

Miners who had once held high hopes for their claims now sold out or gave up. Production declined, and mills, barely able to continue under the best of circumstances, were forced to shut down. This was the first example—but not the last—of overbuilding in the Colorado refining industry.

To make matters worse for those mine owners who continued, water seepage increasingly hampered operations. Some of the mines were unsafe (timbering skills were few) and required improvements before production could be maintained. The costs of mining, fuel, timber, hoisting, and treatment went up meanwhile. Experience with deep mining and ample finances to underwrite development and production were needed before profits would match expectations. Unfortunately, neither was available in Colorado.

At this point, in 1863, with Colorado mining on the brink of becoming the hoax forecast by pessimists in 1859, a savior appeared in the guise of the eastern investor. Back in the States, the Civil War had created a period of industrial boom and inflation in the North. Money was available for investment and speculation, as never before in the country's history. The issuance of paper money seemed to many to be undermining the monetary system. As a hedge against rising costs and a weakened dollar, gold appeared to be the perfect investment. The excitement of 1859 and the continued shipments of gold dust and bullion from Colorado created a natural interest in local mining and mining stocks. Where better to invest than in Gilpin County mines where it had all started?

It was just as natural for Colorado to turn to the East; most of its settlers had come from there. California's investing public, caught up in the Com-

10. In 1864 the New York Gold Mining Company seemed ready to operate on a large scale on what was described as a "handsome ore vein." The neighboring company thought the vein was theirs and a law suit stopped work for years. Lawyers could make fortunes on mining disputes such as this. Courtesy the Denver Public Library, Western History Department.

11. Central City had changed remarkably by the mid 1860s. False-fronted construction predominated. With fine disregard for pollution, outhouses were built right over the creek. By now, settlement had concentrated in a few communities, of which Central was the major one. Courtesy the Denver Public Library, Western History Department.

12. Main Street, Central, 1864, boasted along its wooden sidewalks a hotel, drug store, bookstore, dentist, jeweler, and general merchants. The shopper had a large variety from which to select goods. This was an urban frontier; a farming town of the same age would not have been so blessed. Courtesy the Denver Public Library, Western History Department.

13. The "boys" pose before a famous gambling hall. The band attracted customers and provided entertainment. To the right was the Montana Theater, which in May 1865 featured Colorado favorite Jack Langrishe and his troupe in the "Isle of St. Tropez," and in the laughable farce, "My Son Diana." Courtesy First Federal Savings, Denver.

stock's silver excitement, invested there, not in far away Gilpin County. Throughout its early mining decades, Colorado would look to New York for financial support. The same held true for railroads, mining equipment, and even mining engineers. Of the western mining regions, Colorado was the most eastern oriented. It was also, by the accident of geographical location, the most eastern of the mining areas, except for the Black Hills.

A few sales late in 1863 whetted appetites of investors, who promptly gorged themselves during the winter and spring, pushing Colorado mining into the center ring of American investment. Mines—some only holes in the ground—were sold and traded as producers of the first rank. Coloradans suddenly found the solution to all their problems: sell them to eastern investors and let their money take care of what ailed the mine.

What started as a trickle became a flood; mines that had failed to produce were valued at enormous figures and the gullible fell all over themselves buying stock. So great was the demand that agents were sent to Colorado to hunt up and purchase mining claims. Never had locals seen the like. They responded predictably. Agents readily found claims to purchase. Gilpin, Clear Creek, and Boulder counties all felt the headiness that comes from being courted and wooed. The euphoria did not last. In April 1864, the end came and eastern and Colorado dreams evaporated. Speculators took home fortunes, while stockholders were left with questionable property and stocks. It had been a lark to sell and speculate in stocks, easier than working the mines, but now the time had come to work, if the investments were to be saved. Easterners were learning the lesson, however slowly, that there was a decided difference between gold bullion and gold stock.

Eastern companies rushed headlong into mining, confronting the same problems that had baffled Coloradans. They found profits difficult to come by. Skilled managers were not readily available. So-called "professors," claiming a miracle milling process, cost gullible backers money and patience. And neophyte investors were not particularly wise in spending their

14. Freighting, the lifeline of the plains, was never easy, especially when a boiler was being hauled. Denver was the end of the overland trail in the early 1860s, with Larimer Street looking this way to trail-weary freighters. Courtesy the Denver Public Library, Western History Department.

money. Frank Fossett, an astute and capable reporter, summarized this period: "Meantime very many mining companies had taken steps to work their properties, or at least get rid of their working capital as speedily or foolishly as possible. The entire history of these company investments and operations, with a few exceptions, could hardly have been worse."

Investors and companies got burned, especially when they attempted to find a way to save a higher percentage of gold during the milling process. The most important item was to get the mill built. No time was allowed to determine whether the process would work, or if there was enough ore to justify the building expense. The more novel the milling method, the better it seemed to be accepted. Ore was pulverized, given chemical baths, steamed, and roasted. The final cleanup, the moment of truth, revealed little gold and left behind a trail of frustration and wasted investment. Add to this, incompetent or inattentive supervisory personnel, some cases of dishonesty, and high overhead, and it is easy to understand why most of these companies went bankrupt or simply quit operating. The closing down of so many mines and the failure of eastern investment plunged Colorado into a gloomy mining slump.

Even under the best of circumstances, these eastern companies would have had a hard time earning the profits so essential for the investor back home. Colorado mining was vulnerable because of an immature local economy. Almost every item needed to sustain the industry had to be brought in on the long transportation line stretching from Denver to the Missouri River. All transportation ran at the whim of nature and the Plains Indians, and neither was predictable.

Major trouble finally erupted because of Indians' resentment over encroachment on their land. War broke out in 1863, and the next year the overland trails were closed, for all practical purposes. Trains were stopped or never started, valuable cargoes were lost or delayed, shortages developed, and the cost of living soared. Eggs jumped from 65¢ to $2 a dozen and flour from $16 to $30 a barrel at Central City. Communications, business, day-to-day living were all affected and not until 1865 would the situation change. Even then, costs would stay high for several years. It became so bad that ex-Union General Fitz-John Porter, now a Black Hawk mine manager, wondered: "can some curse be inflicted on this country for some previous sin that everything seems to be attended by bad luck?" Never would Colorado's transportation vulnerability be more clearly demonstrated. Five years would pass before Colorado gained railroad connections, the ultimate answer, and then only after suffering the trauma of being bypassed by the mainline and having to build the connecting tracks into Denver.

While gold grabbed headlines, silver captured adherents in Georgetown and Summit County's Snake River drainage. Reports of silver discoveries had appeared off and on since 1859, with particular excitement becoming evident in 1860. William Byers and his *Rocky Mountain News* had been irrepressible boosters, avidly reporting "silver nuggets" in such places as Tarryall, Buckskin Joe, and California Gulch. However, ore sent east was returned with discouraging reports of silver content; the local "boys" did not have the experience to handle silver, even if they had found it. A miner, Samuel Leach, who saw "silver ore" at Buckskin Joe in 1862, summarized the situation clearly. He noted that it could not be handled locally and went on to say that "it must be sent over the ocean to Wales for treatment and that does not seem worth while."

Interest in silver did not die. In 1864 it was discovered in Clear Creek County, near Georgetown. The next year a "silver mania," as one editor described it, ensued. Claims were staked on every likely outcrop and anxious owners awaited the silver millennium. It did not come. Like gold, the silver ores were decomposed by nature's actions near the surface. Unfortunately, as the miner went deeper, the ores became every bit as refractory as gold, but with a difference —they required even more complicated methods of separation. One process after another was tried, and failed, leaving behind wasted opportunities and money and depression. Colorado's gold and silver mining both anticipated the development of a profitable reduction process. Georgetown became the center for experimentation, and silver mining languished while it awaited the results. Far too much money was being consumed, and too many hopes dashed, by quack processes. One disgruntled observer said the country between Georgetown and Idaho Springs looked like a graveyard of costly monuments: "These 'compliments to the dead' will not be found to resemble marble, nor nothing grand and imperishable, but rather will they be found after the pattern of ruined and ruinous mills, surrounded by old rusty machinery, and decked with scattered fortunes."

Smelting was not a problem that could be solved by a practical or engineering solution; as Rodman Paul recently suggested, what Coloradans needed was "a scientific solution." Fortunately for Colorado mining, Nathaniel Hill arrived on the scene and in 1867 organized his Boston and Colorado Smelting Company. A professor of chemistry (one of the bona fide ones on the Colorado scene), he had been sent out by Boston capitalists in 1864 to investigate their recently purchased mining properties. Hill became intrigued by the ore reduction problem and approached the problem scientifically. He traveled to Britain and Europe to investigate smelting processes, finally sending ores to Swansea, Wales, the world's foremost center. Eventually, Hill contrived a process based on smelting and concentrating gold ore on copper mattes. He opened a smelter in Black Hawk in 1868, a milestone in Colorado mining, because Hill had pointed the way to the solution. Though a technological success, the Hill process remained costly and only high-grade ores could be worked profitably. Hill initially sent the matte to Swansea for final separation, but within seven years localized the entire process, reduced costs, and resolved other problems.

Colorado had suffered through five years of incompetence which, together with the mine and stock speculation, could easily have killed all interest in mining. Fortunately, it survived.

Production figures for individual years during the 1860s are lacking. Generally, after a good start, production slumped badly until 1868. Then it revived. Placer production dropped off steadily after 1862, while lode production rose dramatically. By 1869, when rough estimates are available, eighty percent of the territorial gold came from lode mining. Silver production always lagged behind gold, but picked up as the Georgetown experiments developed a reduction process, based upon amalgamation, similar to what was being used in Nevada and California. It involved crushing, chloridizing, roasting, and amalgamating in pans or barrels and seemed to adapt to most of the district's ores.

Almost unnoticed in the turmoil of the 1860s was the discovery and drilling of a successful oil well near Canon City in 1862. One of the earliest oil wells west of the Mississippi River, it was discovered only three years after the first commercial oil field opened in the United States. Extraction methods were primitive. One well operator bailed his oil out by hand; when quicksand was encountered in several wells, no one knew how to handle it. Isolation, little capital, poor transportation, small market potential, and litigation stopped this enterprise from becoming much more than a curiosity.

One of the crying needs of Colorado had been fuel; easily accessible trees went down under the ax and saw, leaving behind soaring wood prices, as cutters were forced to go farther afield. Exposed coal seams had been noticed almost as early as the arrival of the first pioneers. Nearby residents took advantage of this boon by chopping off enough for home needs. By late 1861 coal was being offered commercially on the Denver market, and by mid-decade mines were operating in the Denver-Golden area. The Belle Monte Furnace Iron and Coal Company, fifteen miles north of Denver, was selling coal in 1866 and, by operating its own blast furnaces, pig iron. Nine-dollars-per-ton freight charges kept the coal price at $16 in Denver. Coal mining also needed better transportation.

The 1860s had been trying years for Colorado mining. Only a few short years separated the excitement of 1859 and the depression of the mid-1860s. The difference between easily worked gold and the difficult refractory ores had often been a matter of only a few feet in the mine. In the short span of a few days, distressed investors went from being paper millionaires to owners of worthless stock. The 1860s was a decade of contrasts—buoyancy and depression, quackery and scientific experimentation, rushes based upon fact and others upon fancy, districts born and districts dead, immigration and emigration.

Those two components for successful mining—sound financial backing and year-round, economical transportation—had come to haunt Colorado. In an industry not noted for its conservatism, it appeared that more money than usual had been wasted. Colorado could ill afford the expense of those failures, or the tarnished reputation it acquired after the gold bubble burst in 1864–65. The investing public turned away from Colorado, now an unwelcome pariah in American mining. A worried Georgetowner wrote in December 1868 that while local people held a high estimate of nearby silver mines, easterners could not "be induced to invest."

Colorado knew what it needed in the way of transportation, it just could not secure the connections. As the decade closed, Denver was close to the goal, but the mining districts would have to wait a few years more before the welcome smoke of the railroad engine sullied their skies. After all the debacles with milling and smelting, Coloradans should have learned a lesson, but each new mining rush was capable of launching another reduction craze based more on enthusiasm than on experience and experimentation.

The plush days of the placer miner came to an end during the 1860s; superseding the individual came corporate mining. The companies remained small, and a lucky individual could still make a fortune, but the odds in his favor were lengthening. The company with money stood to make even more. Lode mining demanded knowledge, experience, and capital that pan and pick placering never did. Colorado mining had not as yet come of age; it limped through a decade marked more by persistence than success. Even the revival of the late 1860s had not convinced some scoffers that the territory would not sink further. Colorado still had to prove itself, and in the western part of Boulder County some Central City men were laying plans to help it do just that.

15. Buckskin Joe had slipped past its peak by the late 1860s and was well on its way to becoming a ghost town. Even at its best, it failed to equal Central City, but the type of architecture and buildings strung out along this dirt street was more typical of the Colorado mining frontier during those years. Courtesy the Library of Congress, Washington.

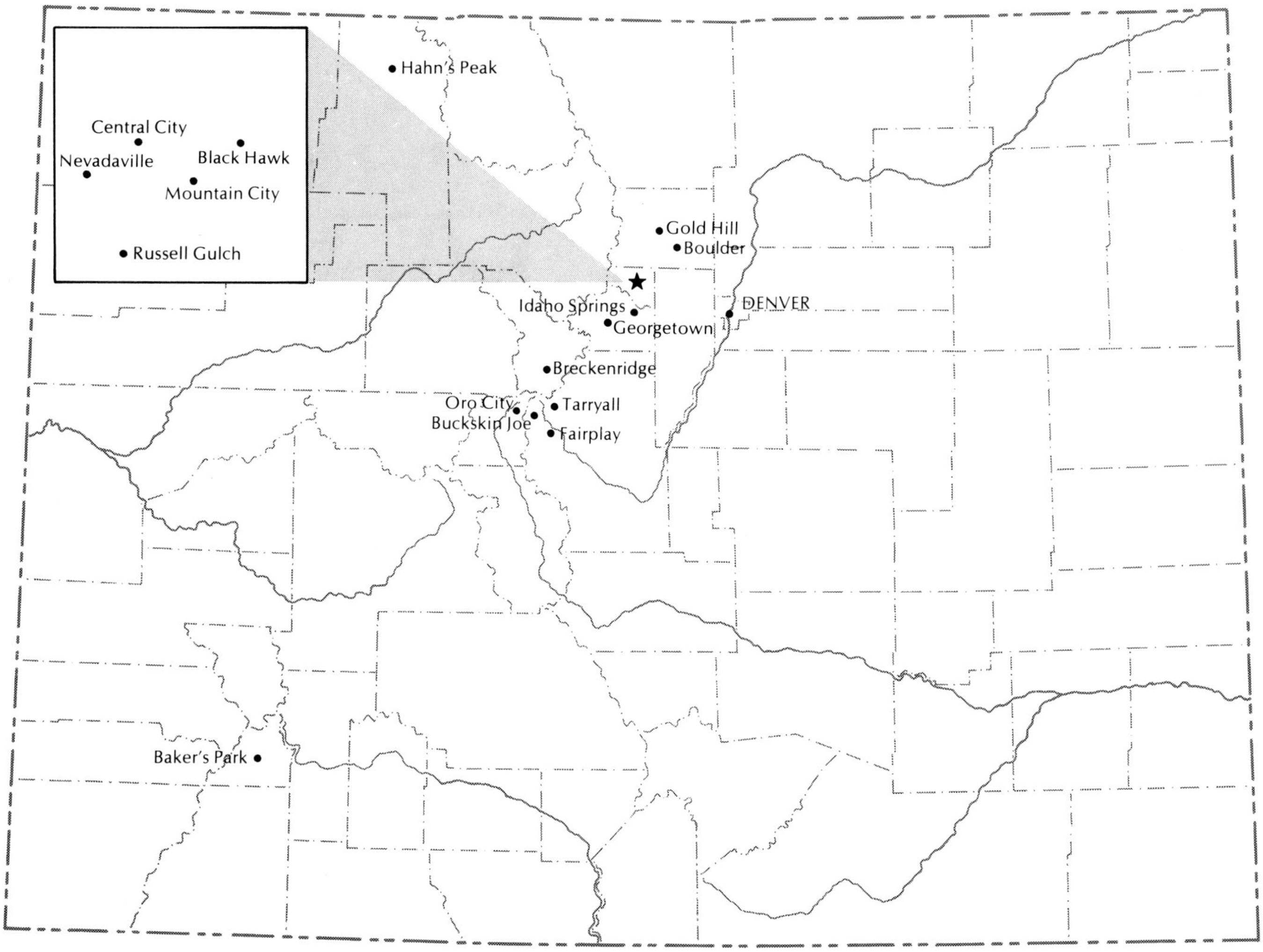

MAP 2 Colorado Mining, 1858–1860s

3

The Silver Decade

The 1870s showered silver upon Colorado in amounts that far exceeded even the most optimistic expectations. It was truly a silver decade. Gold had dominated since the first discoveries; now silver came of age and with it the rise of Colorado to number one mining state in the United States. Silver mining, which had grown slowly during the previous decade, jumped from a production of 496,000 ounces in 1870 to over 11,000,000 nine years later. Where men had once dreamed of thousands of dollars, they now had visions of millions, and the days of 1859, when Leadville became the sensation of the state and nation, returned.

The party of Central City men prospecting within sight of the continental divide in the late summer of 1869 scarcely realized the impact their efforts would have on Colorado. They discovered silver at what became Caribou, and the rush there in the spring inaugurated Colorado's silver decade. Caribou, the most northerly of the state's principal silver districts, was the first of many. As with other districts, its production soon peaked; within a dozen years decline had set in; the drift toward oblivion became inevitable. Before this happened, however, Caribou caught the attention of Colorado. One of its mines sold for an unprecedented three million dollars. Eastern and foreign investors got a taste of investing in Colorado silver mines. Caribou's reduction problems proved minimal; the amalgamation process popular in Georgetown worked well on its ores.

Georgetown—Colorado's first "silver queen," as so many young camps and even mines aspired to call themselves—was the center of a beehive of activity in mining and milling in the early 1870s. The discovery of rich lodes in the surrounding mountains gave it a valid claim to the title, and in 1874, for the first time, it exceeded neighboring Gilpin County in total production. Three years later the railroad chugged into Georgetown, and production for the next seventeen years averaged more than two million dollars per year. Previously that figure had been reached only once. The Summit County (Snake River) silver mines, which had caused almost as much excitement back in the sixties (Hollister laconically called the enthusiasts "snake bitten"), lagged well behind. Served only by wagon roads over high passes, and lacking the mills, resources, or publicity of Georgetown, these mines and their owners could only bide their time. Not until the 1880s would silver mining in Summit County finally come into its own. The contrast between the two areas was all too obvious. Although separated by only a few mountainous miles, they were years removed in acquiring improved transportation and outside financing. Even the resolution of the reduction problems in Georgetown meant little—the ores could not absorb the transportation costs.

Georgetown and Caribou showed that there was money to be made in silver and that reduction problems could be solved. Almost immediately, prospectors swarmed out looking for silver.

Park County, the old placer gold area, had a silver boom centered around Alma. It started in 1871, reached rush dimensions in 1872, and peaked in 1875 with half a million dollars' worth of production. Nathaniel Hill briefly opened a branch smelter in Alma, and soon other smelting and concentrating establishments joined his to handle the increased production.

Park County, for all the interest generated, did not hold the same promise as a brand new mining district, several of which opened in the 1870s. Prospectors and miners once more ventured into the San Juans, now not quite so isolated with settlements ringing the fringes. They were looking for gold, but soon found outcroppings of silver. The goal of their search swiftly changed. The Utes glared threateningly—this was, after all, their reservation—and the government was compelled to negotiate the Brunot Treaty in 1873, whereby the Utes ceded the heart of the mining land, to prevent hostilities and allow prospecting. The "Ute question," as it became known, proved to be only partially resolved; finally, in the 1880s, the Utes were removed from much of the western slope of Colorado to Utah. The impact of mining on the Indians and their way of life was shown more clearly here than anywhere else in Colorado. Nothing would satisfy the miners until the Indians were gone.

Removal of the Utes did not convert the San Juans into a prosperous mining area. A short summer season of work became year-round mining when supporting settlements appeared in 1874–75. The ore was there, without question, but the San Juaners could not make a profit. The cost of living and mining remained high, since all goods had to be freighted in and then repacked on mules and burros to reach the remote mines in the mountains. Some high-grade ore was sent to outside smelters, but most of it did not warrant the shipment costs. Smelters were built at Silverton, Ouray, and Lake City, without much consideration of process, ore type, or ore availability. Several floundered from the start, a few went through a trial-and-error period before righting themselves. Local mining interests, meanwhile, suffered because of high reduction costs and low returns. These costs were further increased since coal and coke had to be freighted to the smelters from as far away as the eastern slope. What the San Juans needed was a railroad to ease the transporation burden; at decade's end it was still several years away.

Nevertheless, the San Juaners were making some progress by advertising their mines and mineral resources. They might have succeeded in finding capital and securing a railroad had attention not been diverted by the cynosure of the decade, Leadville. Everything offered by the San Juans, and for that matter the whole state, paled in comparison. Why invest at Lake City, Alma, or Georgetown when Leadville held out the chance of instant wealth, the likes of which Coloradans had not sampled before.

16. Georgetown became Colorado's first "silver queen" in the mid 1860s. The smoke billowing at the right shows how a smelter or mill could pollute. Add to this coal and wood smoke and it is easy to imagine the pall that often hung over the camps. Of the rapidly mushrooming camps, one eastern correspondent confessed: "It is almost impossible to keep correct our geography or remember the names given." Courtesy the United States Forest Service, Washington.

Oro City, Leadville's older neighbor, had been almost forgotten in the swirl of events since its days of glory. Placer mining had played out generally in California Gulch, to be replaced by lode operations, which failed to generate either territorial curiosity or profits. The site of Oro City had shifted from the lower to the upper end of the gulch to be nearer the mines. This process only accomplished a change in scenery. There was nothing to attract an outside investor or a railroad builder. The Oro City district languished and might have followed many of its neighbors into oblivion. That would not be the case this time.

As far back as 1860, the *Rocky Mountain News* had

17. By the 1870s, Black Hawk was a bustling mining, milling, and business center. It also possessed overhead railroad tracks, a rarity for Colorado. The train reached here in 1872 and brought renewed prosperity. Courtesy the State Historical Society of Colorado, Denver.

18. Nathaniel Hill's Boston and Colorado Smelter was the best in Colorado and the first to achieve technical and financial success. His careful management and skill, and wise selection of supervisory personnel, made it a success. When he moved to Denver, that city took a gigantic step forward as a regional smelting center. Courtesy the Denver Public Library, Western History Department.

19. Hand sorting of ore took a skilled eye; this quiet scene showed the Caribou Mine ore-sorting area. Men who could no longer work underground found this to be an acceptable substitute and it gave boys a good training ground for future mining. Note the lunch pails warming on the stove. From the author's private collection.

20. Water shooting through a high pressure hose and nozzle cut away high tonnages at low cost and allowed low-grade gravel to be worked at this hydraulic operation near Alma. This environmentally disastrous method utilized sluices to recover the gold. The Park County placers were some of the longest-lived in Colorado, yielding to pan, hydraulics, and dredge. Courtesy the Denver Public Library, Western History Department.

21. Mining in the isolated and rugged San Juans was not easy, as the 1870s prospectors and miners found out. This 1875 photograph shows a camp on King Solomon Mountain, near Silverton and its early rival Howardsville, both small hamlets at the moment. Courtesy the United States Geological Survey, Denver and Washington.

22. Colorado's most exciting mining district in the 1870s—Leadville—was centered here on Fryer and Carbonate hills. Horace Tabor made his first million with the Little Pittsburg and Chrysolite mines; others made thousands. Courtesy the Bancroft Library, University of California, Berkeley.

written of silver in California Gulch, and in 1872 the same paper mentioned a silver lode near Oro City. With no local smelters, little available money, and high transportation costs, the Oro citizens could only conjecture about the possibilities, while running samples through the local assay office. Silver and commercially valuable quantities of lead (this was a lead-carbonate area) had been found, and by early 1877 the search carried prospectors slightly northward to a group of hills that would be known individually as Carbonate, Fryer, and Iron. Around their base a new settlement started that summer; before twelve months passed it had completely overwhelmed its neighbor.

Leadville, Colorado's greatest silver camp, opened in 1877–78. By 1880 it was the second largest city in the state, having grown from zero population to over 14,000 by census count, and much higher by local estimates. Instead of production in the hundreds-of-thousands or million-dollar range, Leadville's mines produced $9 million in 1879 and over eleven million the next year. Added to this was a million dollars' worth of lead in 1879, an amount that tripled in 1880. Leadville evolved from mining camp beginnings to mining town status within a year, a goal most camps never achieved in a lifetime. No hard and fast criteria differentiated the two stages; population, production of mines, size of business district, and architectural styles all played roles. But it was obvious to any visitor just which one he found himself in. There was a certain confidence, spirit, even arrogance, that permeated the mining town. It came naturally to the Leadvilles of the mining world; in the less prosperous camps, it appeared forced.

Leadville and its mines yielded both astonishing riches and legends. Its most notable legendary figure was Horace Tabor. After eighteen unspectacular years on the Colorado mining frontier, Tabor moved his general store from Oro City to Leadville and grubstaked his way into wealth and Colorado folklore. He became the symbol of what mining could do for the individual. Leadville meanwhile became the yardstick against which other Colorado camps would be measured. Leadville's red-light district, its rowdiness, its spiciness, its wealth would be talked about for years. Suddenly camps were being referred to as the "new" Leadville and mines were opportunistically named after the big producers that rimmed its northeastern limits.

Leadville became the Mecca for miner, investor, railroadman, merchant, and visitor. Instead of having to woo the railroad, as many camps and districts were forced to do, Leadville had two rail companies fighting for the privilege and profits of serving the town. A census taken in May 1878, when the community was less than a year old, found bakeries, clothing stores, meat markets, drugstores, physicians, banks, dry goods stores, and a newspaper office crowding the main street. People came just to see and sample. A reporter, writing to the *Engineering and Mining Journal* on October 5, 1878, was carried away by what he saw.

> As a glorious camp, rapidly increasing and developing, Leadville is the greatest sensation on record. Its mineral resources are immense, easily developed; communication with the world is laborious and expensive; the climate abominable; . . . the spirit of the population good and happy. Everybody who wants work gets it, and good pay.

Helen Hunt Jackson, novelist and poet, also came in

1878 to enlighten her readers from the woman's viewpoint on the town she called "a marvel." She visited the log cabins, unpainted board shanties, and tents, talking with their residents and coming away convinced that there was no reason that everybody should not be a millionaire. The men, she noticed, had "eager, restless and fierce faces," as they watched the drama of newly made fortunes repeating itself over and over again.

The sensitive writer and artist, Mary Hallock Foote, followed her mining engineer husband to Leadville. She too caught the excitement of 1879: "And all roads lead to Leadville," she wrote a friend about her trip. "Everybody was going there; our fellow citizens as we saw them from the road were more picturesque than pleasing. I was absorbed by this curious exhibition of humanity all along the 70 mile long journey." In subsequent letters she revealed a perception of the unique character of the mining frontier. There was, she believed, a great deal of nonsense and enthusiasm, which could easily have been destroyed, but no one cared to trouble himself about his neighbor's illusions, and everyone was convinced that he himself had none. Each man thought his mine was sure to pay, or if he had not struck it rich, felt quite positively that he would. "The men out here seem such *boys* to me—irrespective of age!" In a real sense they had to remain boys, or succumb to the hard reality and long odds of the mining frontier, whether it was in Leadville or somewhere else.

First and foremost, Leadville symbolized mining—silver mining—and the wealth to be made from it. For the first time in more than ten years eastern money flowed easily into Colorado. No longer did owners or sellers have to seek out and promote their property; investors literally banged on the door to get in on the action. If nothing proved tempting in Leadville, investors traveled throughout the state chasing the rainbow of another Leadville, which they felt was sure to be found.

Fryer Hill, the heart of the 1878–80 bonanza, lodged mines famous throughout the country: the Little Pittsburg produced $375,000 in the first six months of development; the Chrysolite yielded silver ore valued at 187 ounces to the ton; and in January 1880, the Robert E. Lee mined $118,500 in fewer than twenty-four hours. Men could dream, as Mary Foote suggested, when this was reality, and no one could deny that today's hard-luck miner might be tomorrow's wealthy owner.

Both the Little Pittsburg and the Chrysolite were capitalized (the former for $20 million, the latter for $10 million) and listed on the stock market. Promises of dividends stretched ahead for years. Shares, which ranged into the $30-$40 class and were very high for Colorado mining stock of those years, were grabbed up by eager investors. As 1879 closed, dividends did roll in and lucky stockholders from east coast to west sat back in their chairs and watched the promised land of mining investment unfold before their eyes. Denver and Colorado investors, among them Jerome Chaffee, David Moffat, John Routt, and James Grant, purchased properties or dabbled in stock, increasing, or in some cases making, their fortunes and parlaying their wealth into state prominence. Leadville silver stimulated the entire state's economy, nowhere more clearly than in the capital city, where it launched an unmatched decade of expansion and prosperity.

Not only did Leadville provide Colorado with a reputation of success and wealth, but it also prompted tremendously significant strides in smelting. The initial smelters of 1877–78 were based upon testing, and two of the earliest smeltermen, August Meyer and the aforementioned James Grant, were both trained in mining and metallurgy. Smelting was one of the keys to Leadville's good fortune; quite possibly, the mercurial growth and success of that city can be largely attributed to the smelters and the men who ran them. Samuel Emmons, in a report on Leadville smelting published before his more famous study of the region, wrote:

> [lead smelting] has been brought to a relatively high degree of perfection, and is extremely creditable to American metallurgists. One of the most useful practical lessons that has been taught by the comparative success of the various smelting works is that this has been proportional to the more thorough training in scientific metallurgy of its managers, the completeness and accuracy with which they have gauged the operations of their furnaces by chemical tests, and the intelligence with which the results of these tests have been applied to the practical conduct of their business.

Meanwhile, Nathaniel Hill shifted his smelting operations from Black Hawk to Argo, near Denver, in 1878. The reason for this move was economic—it was cheaper to bring ore to a central location than to transport coal, coke, minerals needed for flux (to help promote reduction), and in some cases the ore itself to small, scattered plants tucked away in the mountains. A central location with railroad connections would result in a larger and more modern plant which had the advantage of being able to tap the entire state's and the Rocky Mountains' mineral resources. For the owner it promised more stability and larger profits; for the miner it offered better treatment and perhaps lower rates, at the expense of his long-cherished dream of a local smelter. Hill intended his Argo smelter to be the best in Colorado; his company maintained ore-buying agents and sampling mills around the state and even branched out to work ores or mattes sent from other states and territories. He hired skilled men to work the plant, men like Richard Pearce, one of the best persons in the United States when it came to the metallurgical processes relating to silver. With Hill, Pearce, and the Leadville smelters, Colorado smelting entered a new era; the days of "common sense, fire-brick, ore and wood" were disappearing.

Denver, the railroad hub of Colorado, also became one of its major smelting centers, rivaled for a time by nearby Golden. Both had similar advantages of proximity to coal fields, central locations close to the mountains, and viable wagon and rail transportation routes. No reduction works in the mountains could match these; as a result, the concept of regional smelting centers took root. Mining developed as an industrial complex in Colorado. The individualistic era of the 59'ers receded into memory.

The growth of precious-metals mining, the rise of the smelter industry, and the coming of the railroads stimulated Colorado coal mining. Major mines were found either along the foothills of Boulder and Jefferson counties, or in the foothills in Fremont and Las Animas counties. Frank Fossett estimated production at slightly over 200,000 tons in 1878; it jumped to 327,000 tons in 1879. Boulder and Fremont counties were the vanguard. Trinidad became the center of the coking industry, its coal being the best grade for supplying this vital ingredient to the smelting trade. The first stage of Colorado coal mining, characterized by neighborhood markets and few deep or large-scale operations, ended in the late 1870s and early 1880s.

As coal mining increased in significance and profit potential, interest in it mounted. Very early, 1876–77, reports came from the Animas River Valley in southwestern Colorado that coal had been found there. The pioneering settlers had simply to drive their wagons to the outcroppings and shovel in enough coal for winter's consumption. Local entrepreneurs soon put a stop to this practice by laying claim to the coal seams, but the market was far too small to be profitable at that time. In the newly opening Gunnison country there were reports of coal discoveries, followed by actual mining on a limited scale.

The Gunnison country, like its southwestern neighbor, the San Juans, burst onto the state's mining scene in the 1870s. The parallels between the two are interesting. Gold was discovered in the Gunnison country in 1860, but Indian threats, isolation, and small returns halted active development. Later in the 1860s, summer prospecting uncovered a few silver veins and some coal; however, permanent settlement awaited the 1870s and purchase of Ute land by treaty.

In the 1870s a covey of little camps settled around promising mines; Pitkin, Gothic, Irwin, Tin Cup, and Ohio City were some of them. Gunnison, with the most potential, developed as a business and trading center, much as Boulder had done earlier. A smelter was built at Crested Butte, lacking all but one of the ingredients (coal) previously described as making Hill's operations successful; it did have ready access to coal. This community switched from hard rock to coal mining and prospered. The Leadville bonanza drew attention and settlers away from the Gunnison country; it would have to wait a few more years before the trains came and spurred mining development.

Prospectors moving into the Gunnison district opened the Bonanza silver mines on the western fringe of the San Luis Valley. Soon, four little settlements crowded along the banks of Kerber Creek. Clarence Mayo, who drifted into Bonanza in 1881, wrote his brother that the district was having a "terrible 'boom' and if the mines turn out to be good it will be a 2nd Leadville." They did not turn out that way, and within a month, Mayo went on to Gunnison. Despite pleas from home, Mayo firmly announced, "I will never go home a step till I have made some money." It was this sort of determination, or stubbornness, that helped keep the Colorado mining frontier moving for thirty-five years. More successful was the discovery in 1879 of the silver-lead district on the Roaring Fork River, at the place which later became Aspen. At the moment it was lumped in with the Gunnison country, which, with a few deletions, became Gunnison County.

As the Gunnison experience demonstrates, each rush spawned a host of small mining camps and districts; the same phenomenon was evident in the San Juans, and even the Caribou discovery engendered two small hamlets. Directly around Leadville and born of her fame were Kokomo and Robinson, each with its own aspirations and mines, none of which would equal expectations. More new districts and camps were opened in the 1870s than in any other decade. They appeared anywhere and everywhere, without much forewarning.

Farther south, in the Wet Mountain Valley, Silver Cliff and Rosita, comet-like, blazed in short-lived moments of splendor. Rosita's silver was discovered in 1875, and Silver Cliff came in under the shadow of Leadville in 1878. Production for the county (Custer) —in the seven- to eight-hundred-thousand-dollar range —reached its apex during the next three years, then rapidly declined. Enthusiastic Silver Cliffites, confident that their camp would become the state capital, were doomed to disappointment.

The excitement of the seventies had been matched only once before in Colorado and would be just once again, though never on a statewide basis. It seemed

23. Silver Cliff had high hopes and a period of frantic activity before its demise. Mining camps, even in such an open space as this, were crowded together; spacious lots and lawns were ignored in the scramble for wealth. Courtesy First Federal Savings, Denver.

that all of the high mountains were one vast silver mine. To a degree they were silver-bearing from Caribou to Rico, in a bow stretching southwestwardly. According to legend, the prospector and his trusty mule had opened up many of these discoveries. Perhaps they had, but corporations quickly gained control of the best properties and many of the rest. Men with money stood to rake in more; mining had become a big business that required capital and training—muscle and hope were not enough. The days of the mine owner working beside his hired hands were disappearing, or were already gone, as in Leadville. The practice of working for a while, then striking out on one's own, gave way to absentee ownership and large-scale operations that were well beyond the means of the average individual, even in company with his friends. The ramifications of this development were not as yet clearly perceived.

Scientific mining and milling made great strides. Suspicion and distrust of trained mining engineers by graduates of the school of hard knocks were lessening. Leadville's mining district became a training ground, not only for men, but for equipment as well. The money was available, and the stockholders were usually willing to try something different, if it promised better returns. It was possible that they had no idea how their money was being spent. Their main concern was that the dividends keep coming. Improved and larger hoisting machinery, machine drills and steam engines, grander and more complicated mining methods, and some unrelated inventions, such as the telephone, came to Leadville. Methods and machinery perfected in Leadville found ready acceptance in Gunnison, Boulder, and other mining areas. The vast contrast between Central City in 1859 and Leadville in 1879 was based on more than just the passing of two decades. A comparison of transportation networks, financial support, mining technology, and scientific methodology revealed just how far Colorado had come.

Silver was queen, but gold mining persisted. Boulder County went through a series of small rushes in the 1870s, basically gold-related; such old camps as Gold Hill and Jamestown revived, and Sunshine and Magnolia were born. Their individual fame was fleeting. Henry Wood, a miner who lived in Sunshine in 1877, remembered it as a place where "most of the people were awaiting as I was for some new mining field to be opened." Only the lateness of the season kept Wood from going to Leadville as soon as he heard of the silver discoveries. He stayed until 1878, then deserted stagnating Sunshine for booming Leadville. The assay

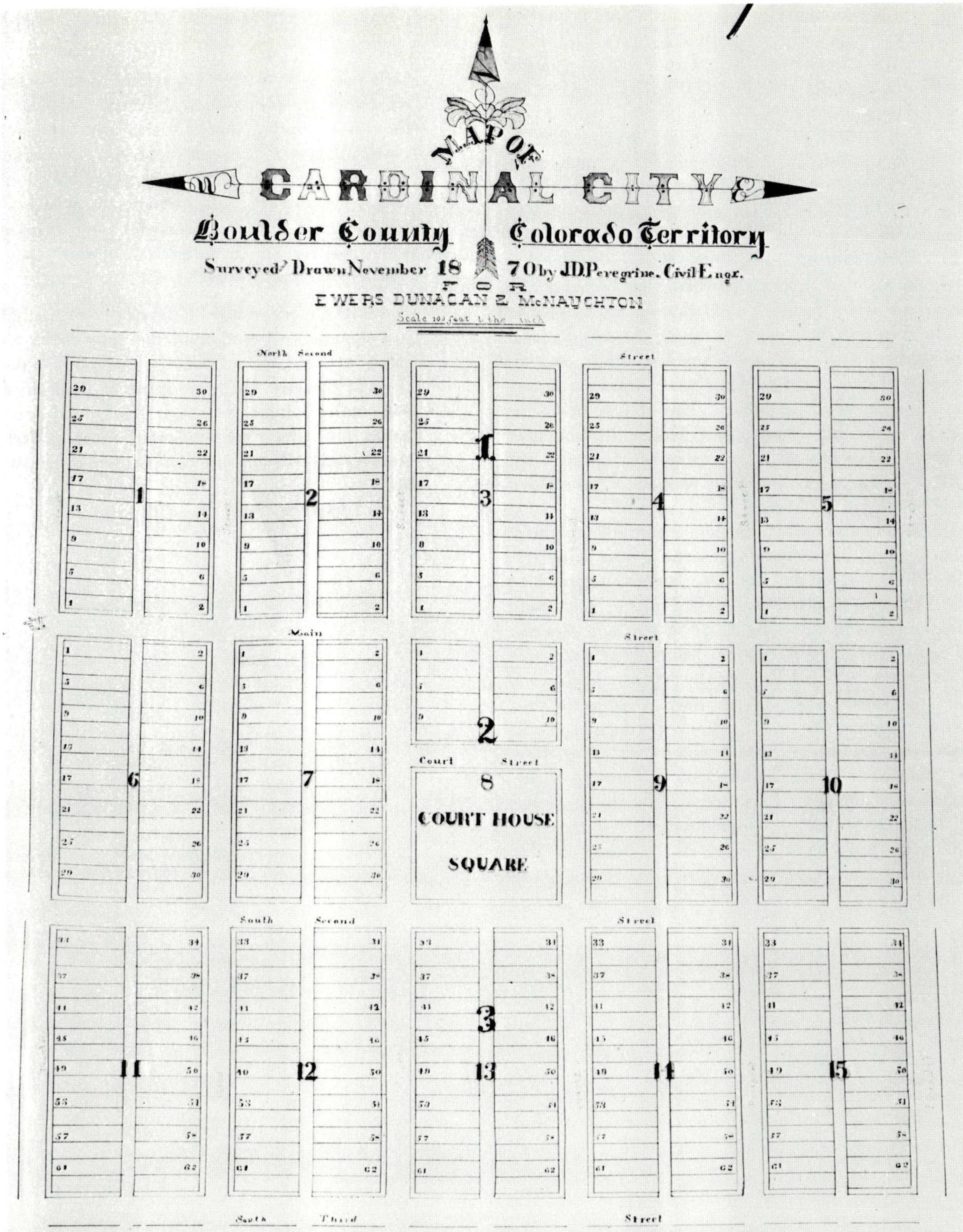

24. For every Leadville or Silver Cliff, there was a score of camps that died in birth or failed to develop fully. Cardinal was never a city, never had a courthouse, and never filled more than a tiny number of the platted lots. Attaching "city" to a camp's name gave it at least the semblance of grandeur. Courtesy the Western Historical Collections, University of Colorado, Boulder.

office he opened there put him on his way to becoming one of the state's better known assayers. All together, these little camps and districts pushed the county's gold production to over $400,000 in 1877, the best up to that date for Boulder.

Gilpin County easily topped that with production that reached a decade high of over $3 million in 1871 (the greatest single year in the county's history). The nearness of Hill's smelter and the railroad, which finally reached Black Hawk and Central City in 1877–78, help explain this figure, as does the fact that the miners and millmen learned how to work the district's ore. Although Gilpin County lost miners to almost every mining madness during the decade, local production stabilized and gained a much sounder footing than ever before when more steam machinery and pumps were introduced. The "old reliable," as some called the county, had been a leading gold district for twenty years and its camps, Central City, Black Hawk, and Nevadaville, were among the oldest still active in Colorado.

Colorado also had at this time a producing oil field. The wells of the 1860s, which yielded some primitively refined fuel oil and kerosene, were superseded by development in what became the Florence field, south of the town of the same name in Fremont County. Lack of capital continued to hamper the two small companies in the area, as did the expensive failures of dry wells, some over 900 feet deep. Despite the slow start, Florence was the state's only producing oil field until past the turn of the century.

The silver seventies had been an exhilarating experience for Colorado. Excitement, adventure, profits, growth, statewide development had resulted in a bonanza. And the future dazzled the beholder. Two quotes from the era summarize the climate of the 1870s well.

> Men pass here for what they are, and not for what they have, how they are dressed or where they were born. Nobody cares who a man's grandfather was or of what state he is a native. No one can afford to treat another with contempt because he is unfortunate; the wheel-of-fortune may turn over and the poor man of today may become the millionaire of tomorrow.
> [*New York Tribune*, July 7, 1879]

> In short, the era of mining development has but just commenced, and with the indications already at hand we confidently predict that the close of the year 1878 will bring to the light a bonanza of mineral wealth in [Colorado] compared with which the celebrated mines of Utah and Nevada will sink into comparative insignificance.
> [*New York Times*, May 20, 1878]

TABLE 1
Colorado Gold and Silver Production
1858–1880

	Gold	*Silver*
1858–69	$30,211,784	$ 1,302,289
1870–75	15,691,400	11,595,916
1876–80	15,561,381	36,416,359
PEAK PRODUCTION YEARS		
Gold	1900–1905	$155,000,000
Silver	1889–1894	113,000,000

Source: Henderson, *Mining in Colorado*

4

To Capture a Shadow—A Photographic Essay

In *A Christmas Carol* Charles Dickens has the ghost of Christmas past transporting Scrooge back through the years to see things as they once had been. To his wondering and ill-at-ease companion, the ghost turned and said, "These are but shadows of the things that have been. They have no consciousness of us." This brief comment well describes the pages that follow.

They are designed to turn back the fading years to the time of the mining camps and towns, not only to show buildings and mines, but also people. For a moment time is frozen, and people are seen going about their everyday lives or posing for a formal portrait. One must observe carefully, for this light's pale beam is all that illuminates people and memories that have long since vanished like last night's dreams.

The sounds and smells are masked—that part of their world cannot ever be recovered. These people and places are not to be looked upon with nostalgia, laced with sentimentality; they deserve better. They did not romanticize what they were doing, and there is no reason to do it now. They once had their own hopes and expectations, joys and disappointments. As they stare out from years past, it is haunting not to know what they are thinking, what they did in the next hour, the next week. Yet for a moment they beckon us to come back to see, to gain a richer understanding of their generation.

Near the end of his journey into the past, Scrooge asked to be removed; it overwhelmed him, proved too haunting. We must avoid that trap, examine carefully what follows, and try to capture a shadow. Because, as the ghost observed, "I told you these were shadows of the things that have been. That they are what they are, do not blame me!"

25. Present in virtually all camps were the merchants, from the general storekeeper to the specialized shopkeeper of the large towns. The clerks were ready and waiting in this well-supplied Ouray dry goods store. From the author's private collection.

26. The fifty-niners ate the cattle that hauled them to the mines; later, shoppers could go to the meat market and select their cuts. These Nevadaville (the town's official name was Bald Mountain) butchers seem determined to serve their customers. Courtesy the Denver Public Library, Western History Department.

27. This Cripple Creek grocery store was a direct descendant of earlier general stores which had carried a bit of everything. One wonders where the customers were at 3:45. Heinz believed in advertising, and, with a magnifying glass, one can spot other companies, including Quaker Oats, Libbys, and Bakers. Courtesy the Cripple Creek District Museum, Inc., Cripple Creek.

28. A social center of any camp worth its salt was the hotel. Here visitors were lodged, and residents could sit and talk about mining or the current gossip, or simply relax. Courtesy First Federal Savings, Denver.

29. A camp without a newspaper was adrift. No other agency better promoted, advertised, or served as a general gadfly. This was one of Silverton's fanciest buildings in the early 1880s. Next door, to the left, were a surveyor and an assayer, both important in the mining community. Courtesy the Center of Southwest Studies, Fort Lewis College, Durango.

30. One wag said you could not throw a stick across Chestnut Avenue in Leadville without hitting several lawyers. They flocked to the larger communities and became a pillar of the professional community. There was much work to be done and money to be made; in fact, lawyers often had the best incomes in town. Courtesy First Federal Savings, Denver.

31. Each camp had its private entrepreneurs. This fellow provided Nevadaville's water supply in 1911. Getting pure water and maintaining sanitation were two serious problems, and the number of epidemics testified to the failure to resolve them. Courtesy the Denver Public Library, Western History Department.

32. Around the camps and in the hills mushroomed miners' cabins. These three men appear rather well dressed for the meal they had prepared. Home cooking, such as this, was edible for a while, but one reason so many men eagerly anticipated a trip to town was to get a real "home cooked" meal. Courtesy the Denver Public Library, Western History Department.

33. Relaxing amid a collection of pinups and nature scenes, this miner appears quite relaxed with his book. More typically, a lamp, stove, bed, chairs, pans, and eating utensils sufficed as furnishings. His dog looks as though he would rather be outside. Courtesy the Denver Public Library, Western History Department.

34. Prostitutes were found throughout Colorado mining regions. Usually relegated to one street or section of the community, the "fair but frail" worked out of cribs and parlor houses such as this one, reportedly in Leadville. Theirs was not a life of glamour, as legend often states: the startlingly high number of suicides among the "erring sisters" speaks volumes. Courtesy the Denver Public Library, Western History Department.

35. Dance halls were part of the red-light district, as were saloons, variety theaters, gambling halls, and the "girls," but these dancers were not necessarily prostitutes. The band in the background is ready to play, the men have their favorites, and the dance will begin as soon as the photographer finishes. Courtesy the Colorado College Library, Colorado Room, Colorado Springs.

36. The "boys" came to town to relax and have fun and they found the opportunities. These four look quite refined, but ready to go, in their formal portrait. One wonders who owned the dogs. Courtesy the Western Historical Collections, University of Colorado, Boulder.

37. Their twentieth-century descendants might have more of a selection of photographs, but the nineteenth-century miners were not remiss. From the author's private collection.

38. "Bunny" on a mule. From the author's private collection.

39. Nineteenth-century "bunnies" graced the walls of many a miner's cabin. From the author's private collection.

40. The ubiquitous saloon—no camp was complete without at least one. It was a man's castle, where he could go for a beer, companionship, a game, or business. This Telluride establishment came complete with nudes, roulette, brass spittoons, and a relaxing lawman. Courtesy Mrs. Homer Reid, Telluride.

41. If the "boys" or "girls" got too far out of line, the constable was called in to restore order. This Cripple Creek force was much larger than the average, which normally consisted of one day and one night man. The camps were never as lawless as the westerns would have us believe; corralling drunks and stray dogs was far more typical than a gunfight. Courtesy First Federal Savings, Denver.

42. When wives came west with children, the days of the wide-open mining camps were numbered. The wedding ring and broom tamed the camps faster than the badge or gun. The women wanted their community to be "civilized" and, in their own way and time, brought civilization about. They wanted a husband home at night, the streets safe for respectable women, and a home that lived up to its name. Courtesy the Western Historical Collections, University of Colorado, Boulder.

43. The homes they had varied: this log cabin was typical for a new district. A wife and mother, however, gave it a touch that no bachelor could. It was not long before frame houses replaced the log cabins. Courtesy the Colorado College Library, Colorado Room, Colorado Springs.

44. The parlor of this Cripple Creek home shows that a prosperous mining family enjoyed the comforts of life. Books, piano, rugs, and electric lights indicate that this mine manager was succeeding in his chosen field. Such affluence was not typical of the men working for him. Courtesy the Cripple Creek District Museum, Inc., Cripple Creek.

45. River City, of "Music Man" fame, would have been shocked to see women playing pool. This was Ladies' Day at the YMCA in Cameron, and the ladies were taking advantage of it. Many of the coal camps had Y's to give the men a choice of recreation. Courtesy the State Historical Society of Colorado, Denver.

46. Women in the mines were believed to be bad luck, but enough photographs exist to show that they did tour some of the mines in the nineteenth century. This group appears scared to death. Or perhaps they were simply frozen by the photographer's flash. Courtesy the University of Colorado Museum, Boulder.

47. One of the first things Mother wanted was to send Billy or Jane to school. She was backed by the businessmen and other civic boosters who felt a school would bolster the camp's image. The teachers never received much pay, and the school term might run anywhere from three to nine months. This was a woman's occupation although a few men were found teaching. The young scholars in the photograph were Leadvillites. Courtesy the State Historical Society of Colorado, Denver.

48. Playgrounds were much more fun than classrooms. This one, located in Lime, shows that not all of the company towns were dirty, ugly communities. Children matured rapidly—despite curfews and ordinances they could see all sides of life at an early age. Boys went to work in their teens and, with the shortage of women, few girls reached twenty still single. Courtesy the State Historical Society of Colorado, Denver.

49. Christmas and the Fourth of July were the two major holidays. This Gold Hill group was celebrating the latter, although those two boys in front would most likely have had more fun dressed in less "respectable" outfits. Courtesy First Federal Savings, Denver.

50. Mining camp residents danced away many an evening, especially during the winter months. Here was a chance to get dressed up, court one's best beau, and have a midnight supper, all for the price of a ticket. Courtesy the Boulder Historical Society, Boulder.

51. This Telluride group was planning a "mule skinners" ball; despite the apparent drinking, it was a sober affair. Rowdiness was not condoned. From the author's private collection.

52. Like the school, church, and courthouse, an opera house was beneficial to the civic image, and this one actually presented operas. Leadville's Tabor Opera House was an example of the finest. Horace Tabor became one of the earliest Colorado bonanza kings to reinvest his fortune at home. Courtesy the Henry E. Huntington Library, San Marino, California.

53. A source of local pride and an outlet for individual talents was the band. Attired in their uniforms, these Red Cliff band members appear ready to march. Courtesy the State Historical Society of Colorado, Denver.

54. Some camps even had orchestras, or so they were called. This White Pine group featured banjos and horns. Dances and occasional concerts filled their performance dates. Courtesy the State Historical Society of Colorado, Denver.

55. Hose races were fun and practical as well. It was important to be able to reach a fire quickly and get set up, which was what these teams were racing to do. The urge to gamble was not unassuaged either; money changed hands on the outcome. Courtesy the Colorado College Library, Colorado Room, Colorado Springs.

56. Baseball was truly the national pastime and few camps were so remiss that they did not have a team. A bigger town like Leadville might import the best players available. This was the Silver Plume baseball team in 1889. Courtesy the Denver Public Library, Western History Department.

57. Skiing (it was first called snow-shoeing) was utilitarian and fun. The skis were long, the bindings primitive, and one pole served to steady the skier. The outfits differed from current fashions. Before lifts and tailored runs, a certain adventure and vigorous workout awaited those who ventured onto the slopes. From the author's private collection.

58. The ranks of the Grand Army of the Republic thinned as the years passed, but they could be counted on to turn out for a parade, particularly on Memorial Day. Aspen was a young camp when this photograph was taken. Courtesy the Western Historical Collections, University of Colorado, Boulder.

59. The church and its membership played a significant role in smoothing the rough edges of camp life. Women were the backbone of religious life; church provided one of their primary social outlets. The building was important: it gave class to the skyline. These Catholics proudly gathered in front of their new church in Ward. Courtesy the Boulder Historical Society, Boulder.

60. Many camps never acquired a church building, remaining just a stop on the circuit. Ministers held meetings in places ranging from saloons to schools. If they were willing to adjust to conditions as they found them, members could be recruited; if not, it was best to move on. This tent was the forerunner of a more permanent Cripple Creek church. Courtesy the State Historical Society of Colorado, Denver.

61. Death was a daily companion. Mining was and is a dangerous occupation, and the camps were not paragons of sanitation. These factors, plus high elevation and harsh climate, made even the common cold dangerous. This funeral took place in Central City. Courtesy the Denver Public Library, Western History Department.

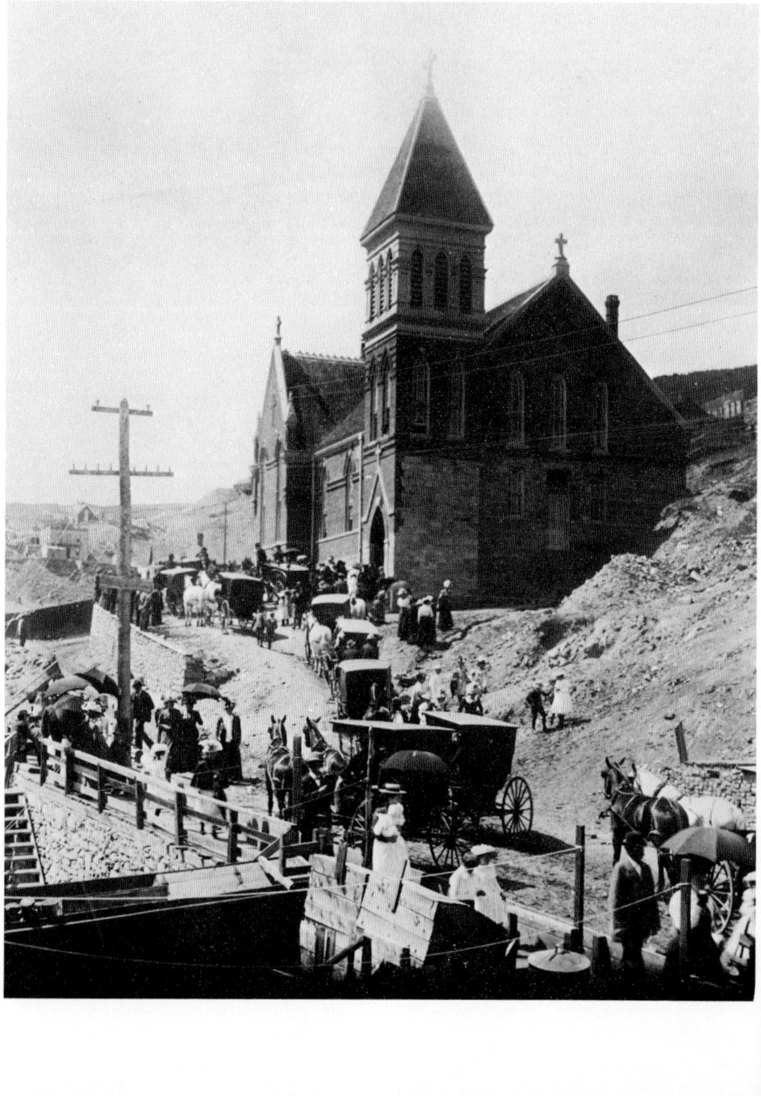

62. A coal mining disaster at Starkville in 1901 widowed and orphaned this group. Most, if not all, were Polish immigrants. Mining went on and so did Starkville, but for these people it would never be the same again. Courtesy the Denver Public Library, Western History Department.

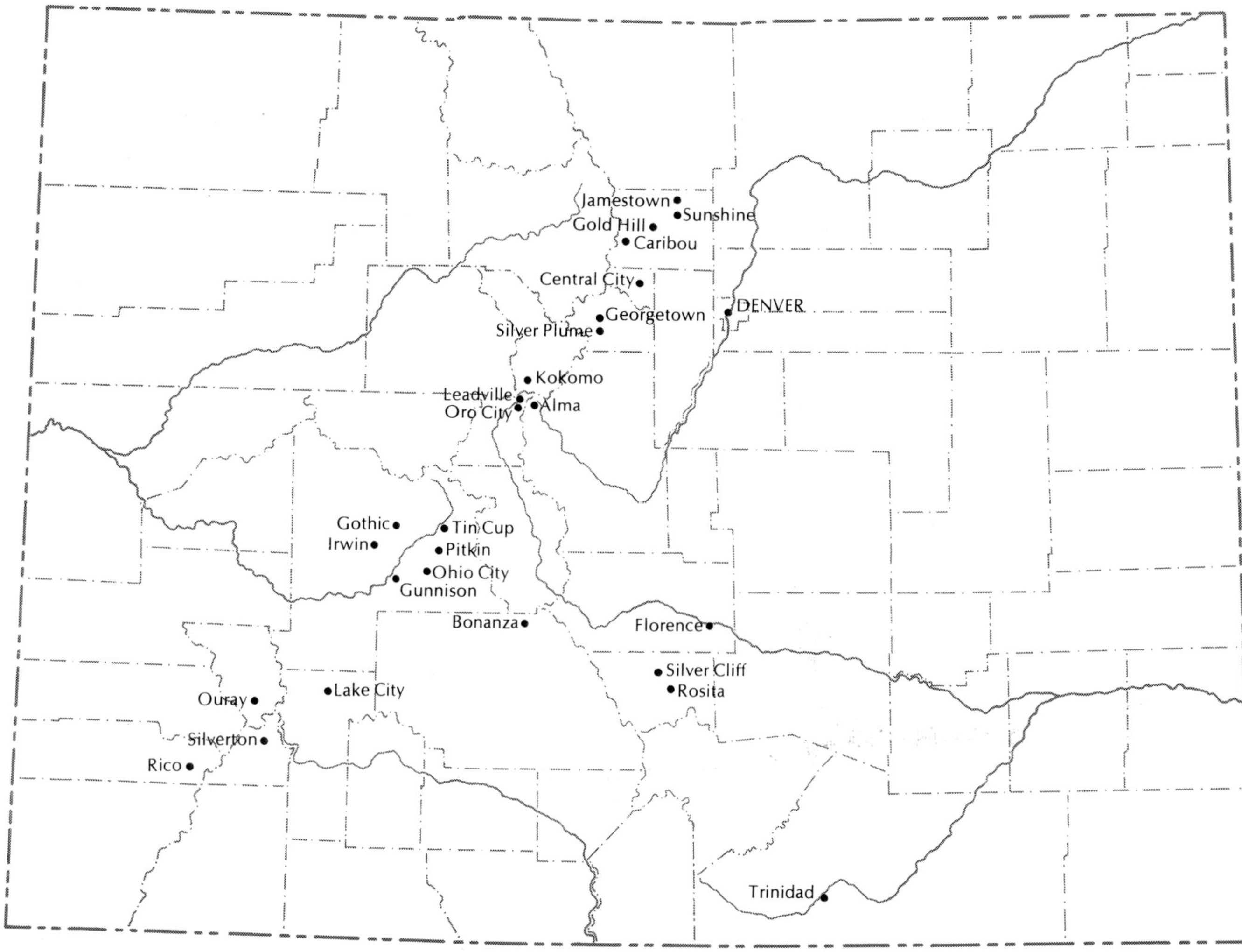

MAP 3 Colorado Mining, 1870s

5

We're Number One

Silver and Colorado were nearly synonymous in the 1880s. Colorado gained and held the nation's number one rank in mineral production. Silver mining was primarily responsible, although gold maintained a second or third position nationally. Gold, in fact, held steady with a three- to four-million-dollar annual production. Silver ranged from a low of eleven to a high of seventeen million dollars.

To the investing public Colorado mining meant Leadville and its silver mines. Those two idols of every armchair stockholder, high production and dividends, roared into the 1880s on what appeared to be a clear track to the promised land of easy wealth. Leadville's two glamour mines performed just as their promoters had promised. The Little Pittsburg paid seven monthly dividends of $100,000 each from June through December 1879, and by mid-January 1880, $850,000 had been disbursed to delighted stockholders. The Chrysolite was hardly less bountiful and declared a $200,000 dividend in February. Stock prices soared, as investors clamored to buy.

Then came the day of reckoning. During February and March Little Pittsburg production faltered and failed; the Chrysolite took a similar path from July to September. Stock prices plummeted, dividends were passed, profits disappeared, and both companies collapsed. Eastern stockholders blamed western management and manipulation; Coloradans blamed eastern greed and stupidity. Each side commanded enough truth to make it self-righteous. Management had gutted the mines to produce those cherished dividends—at the expense of careful exploration and development. Manipulation of stock prices and unfair, if not fraudulent, use of inside information had allowed certain individuals to dump stock on the unsuspecting public. Capitalization and expectations that were both too high led to troubles once the richest and most easily mined ore was passed. Leadville's stocks, once the darlings of the market, now found themselves pariahs, and they dragged with them the stock of most of Colorado's well-known mines. It would be several years before a recovery could be sustained, but never again with the same carefree naïveté of 1879.

These two blows staggered Leadville and Colorado but did not put an end to the troubles that year. In May the first major labor dispute in the Colorado hardrock mining industry erupted. Chrysolite miners walked off the job and the strike spread to other nearby mines. The causes mystified contemporaries and are still unclear today. The miners talked about higher wages, irritating work rules, and management's discharging some men without justification. It is more possible, however, that the Chrysolite company willfully incited rebellion by toughening policies in order to give themselves a rational excuse to pass a dividend, thus buying some desperately needed time. For whatever cause, the strike took a course that foreshadowed others. Threats of violence, action and reaction, heated words, sensational stories, biased reporting, and hardening of positions followed the May walkout. Pressure was finally exerted on Governor Frederick Pitkin and on June 13 he declared martial law and sent in the Colorado National Guard to prevent violence. The strike was broken, the owners emerged triumphant. There had been a weak Miners' Cooperative Union at Leadville, though it had done little to initiate the issues and had played a minor role throughout. Following this strike and the one-sided victory, Colorado hardrock mining was peaceful (except for a few scattered strikes) throughout the remainder of the decade.

Leadville's mines, now labeled "notorious" rather than "eminent," dragged on the market. Management and labor squabbled openly, but corporation rule was firmly enforced, the first fruit of the changing Colorado mining situation. Troops had garrisoned a mining town, Leadville, and its mines. That certain spark that distinguished a vigorous, youthful district from one slipping into middle age flickered out in 1880, the year of tribulation.

Leadville stabilized and then began to decline; both its population and silver values were down significantly by 1889. Only lead production maintained a constant level during the decade and, while gold increased, only twice did it top half a million dollars. Paralleling the silver slump, the local smelting industry, once the best in Colorado, found its relative position declining in relation to the rest of the state, as new smelting centers appeared. The bonanza mines of Fryer Hill were replaced by others on Iron Hill and elsewhere; never, though, was the excitement or interest of the seventies revived. Hope for a second contact or another ore bed somewhere deeper in the earth spurred exploration in the older mines and in some of the newer ones, with mixed results. As the years passed, many of the properties were leased to small operators in return for a percentage of royalty of production. Leadville newspapers interpreted this as a good sign that mining would be encouraged, which it probably was on a short-term basis. But, in all of Colorado history, very few mines were leased when they had high-grade ore or even when there were hopes of finding it. The dumps of some of the big producers were also being worked over to salvage ore previously tossed there as too low grade.

Interestingly enough, Leadville's ore production was not down as much as was ore *value*. The problem was the declining price per ounce of silver, which dropped from $1.15 (1880) to 94¢ (1889); lead slipped also, to less than 4¢ a pound in 1889, a penny lower than 1880. Contrasting this with the higher costs of deeper mining and the lower grade ore brings into sharper focus the troubles of Leadville and most of the older silver districts. This was the start of what was known as the "silver question," which would agitate the entire state in the 1890s. When one considers the odds against success in mining (one study found that 19,499 claims had been filed in the Leadville district, of which 166 produced ore, and only 85 were said to have paid any real profit), it is understandable that the owners sought to avoid lengthening those odds any more by allowing the cost of silver to drop while costs spiraled.

While the old queen suffered through the agonies of her lost youth, she was being challenged by upstart Aspen and the late-blooming San Juans. The camp of Ashcroft caught prospectors' and miners' attention first; then, by the mid-eighties, interest shifted to Aspen some twelve miles away. The great drawback, here as elsewhere, was isolation, the nearest railroad being forty miles away over mountainous roads and high passes. Only the richest ore could be profitably mined. Fortunately, Aspen's potential aroused enough interest to encourage a race between the Colorado Midland and the Denver and Rio Grande railways to get there. The D&RG won by several months in October 1887, and production, which had only once in eight years topped $1 million, was four times that in 1888 and would nearly equal Leadville's the next year.

Aspen was plagued by more than transportation worries; some of its important mines were tied up or harassed by lawsuits for years. This curse hit nearly all mining districts sooner or later. Unclear claim markers, misplaced record books, disputed discovery or ownership rights, overlapping claims, attempts to simply weasel in on a rich mine, and sundry criminal violations threw owners and companies into court. Nothing could have been more exasperating or time consuming than litigation involving the apex issue. An owner had the exclusive right to mine a vein throughout its entire depth and length, if it had a top or apex within the boundary lines of his claim. He could follow it anywhere. This created trouble when the location of the apex was in question. If the issue could not be settled out of court (which it often was in order to save expense and fuss), the trials became complicated arenas of conflicting testimony, exhibits, expert witnesses, and lawyers' skills. A company or individual could be bled white before it was over, and the victor might be little better off than the vanquished. Lawyers made fortunes in mining cases; a rich mine owner and speculator like Horace Tabor kept several on his payroll just to represent him in court. Generally, the richer the mine or district, the more vulturelike the lawsuits that would circle it.

63. The Robert E. Lee Mine rose from "promising" in the 1870s to "bonanza." In January 1880 it produced $118,500 of silver in a twenty-four-hour period, an astounding amount for that time. Leadville, for the moment, basked in its glory, but the Robert E. Lee was finished as a big producer within the year. Courtesy the Lake County Civic Center Association, Leadville.

64. Rivals to Leadville soon emerged. Rico, in the far southwestern end of the Colorado mining belt, proclaimed itself a "second Leadville." Its isolation made it necessary for teams to carry supplies long after trains had landed them elsewhere. Rico never lived up to its expectations. Courtesy the State Historical Society of Colorado, Denver.

65. A more lasting threat to Leadville than Rico was Aspen, which rushed into bonanza in the late 1880s. The burro was a familiar figure in the camps, keeping residents awake with his serenading and every bit as stubborn as the legends make him. The animal was invaluable to miners despite his idiosyncrasies. Courtesy the State Historical Society of Colorado, Denver.

66. Mining was becoming industrialized. Durango, founded in 1880, had a little mining nearby, but was better known as a regional smelter, trade, and transportation center. The San Juan and New York smelter stands in the foreground. From the author's private collection.

Meanwhile, its mines in rich ore, Aspen took on the airs of a mining town, with a population officially over 5,000 and such attainments as an opera house, a fancy hotel, telephones, a waterworks, a daily newspaper, and "all modern conveniences." A local booster was enthralled enough to say, "No town of its size in the state can boast of a finer class of inhabitants or better society."

The editor of Aspen's *Rocky Mountain Sun*, on October 3, 1885, observed more practically: "For three or four years the ranchers of the Roaring Fork valley and its tributaries had to work against vast and disheartening odds. This year they are so well getting on their feet, that they begin to make money." It was axiomatic that wherever the miners went a profitable market would develop for the farmer and the rancher. Where it was possible in the mining country, such as around Durango and Gunnison and along the eastern foothills of Boulder County, agriculture took root in response to the miners' lack of self-sufficiency. Though its market depended entirely on mining, local agriculture established itself so firmly that it provided Colorado with a durable, major pillar for its economic system, one which survived the vicissitudes of mining.

Unlike Leadville and Aspen districts, both limited in size, the San Juans were a conglomeration of many districts loosely tied together by virtue of being in the San Juan Mountains. This area was generally defined by a line drawn from Lake City to Ouray, Telluride, Rico, Durango, and around to what would become Creede. A few other areas, such as Summitville near Del Norte, were included for convenience. Like Aspen, they failed to prosper until the railroad came, and for them the wait was longer. Silverton got an extension from Durango in 1882, and the others finally acquired one in the late 1880s. Needless to say, production improved as soon as rail connections were secured.

Because of their sheer size and the variety of minerals they contained, the San Juans were to have a long mining history. Of all the major Colorado mining regions, this was the least known to the general public, perhaps because of its isolation and slow growth. There were intraregional rushes, which took the "boys" from Lake City to Silverton and to all the major camps, as well as to a legion of smaller hamlets. The San Juans also went through a smelter craze, when almost every district thought it needed one, if not several, smelters. Eventually Durango, blessed with good railroad connections and nearby coal deposits, emerged as the regional smelting center, a position it held for decades.

Great strides were made in production in the 1880s. Ouray, San Juan, and San Miguel counties—the big three—topped the million-dollar plateau and Dolores County was not far behind. The San Juans, however, remained first in the shadow of Leadville and then of Aspen, which stole headlines and investors' attention. Of long-range importance, however, was the fact that the ratio between gold and silver production in most of the San Juan districts was much closer than in other Colorado districts. The difficult conditions of cloud-touching elevations, mountain-backed isolation, and costly four-footed transportation encouraged San

Juaners to experiment early with tramways that stretched down and around mountainsides from mine to mill or to railroad tracks. With coal fuel expenses high and suppliers liable to be detained by winter storms, a few innovative men were discussing and experimenting with electricity by the close of the eighties. San Juan miners, like those in other areas, were agreeable to change if it could be shown that money was to be saved—or made—in the process.

Gunnison County failed to live up to its supporters' expectations, even with the arrival of the Denver and Rio Grande to Gunnison and Crested Butte in 1881–82. The best years were those immediately after the arrival, when such districts as Ruby and Tin Cup were mining lead-silver ore. In the eastern part, Ohio City and Pitkin produced much of this region's gold, but by mid-decade production value was noticeably down and would remain so, except for one brief flurry in the 1890s.

The history of the Gunnison country can be viewed as a microcosm of Colorado's entire mining frontier. One can witness the birth, decline, and death of evanescent mining camps, some in strikingly beautiful locations, each called home by people who spent time there working and just living. The same joys and sorrows that came to all mining communities, to all families, came to Gunnison. Some outside capital flowed in for developing a few large mines which dominated their small districts, creating local notoriety and making a brief mark on the state scene. Within ten years or less, the excitement and the attention had gone elsewhere, and nature began to reclaim her own. Mining people strove for a great deal, but inevitably they were forced to accept less. Mining was and is an exploitative industry; once the minerals are harvested, they can never be replanted. The ore deposits proved too shallow and too poor, or else the veins pinched out too soon, to maintain the districts and keep the Gunnison country in the forefront of Colorado mining.

Some of the older districts were doing much better than the young upstarts: Gilpin and Clear Creek counties were completing their third decade of mining; Black Hawk-Central City-Nevadaville had been mined hard starting in 1859. A steady gold contributor in the million-dollar range since 1867, Gilpin County produced well throughout the 1880s, also contributing between one and two hundred thousand dollars' worth of silver each year. It easily maintained its ranking as Colorado's number one gold district. Those heady forecasts of 1859 were coming true, although the "boys" of '59 were nearly all gone, and pans and sluices and individualism had been replaced by drills and powder and company control.

Life in the camps and towns settled down as well.

67. Miners were working men by this time and held little hope of becoming owners through prospecting. This crew was off to work at the Saratoga Mine near Central City. Cages such as this replaced the bucket hoist by the 1880s in major mines, and steel cable replaced hemp. Courtesy the Denver Public Library, Western History Department.

Many of the smaller ones had long since become relics of another age. Those who did not go to new districts moved into the nearby communities, helping them to avoid a noticeable decline in population. Life took on a more routine existence, and the hustle and bustle, the rapid changes, the verve of the boom days disappeared. Churches, schools, a stable business district, and the gentility of Victorian America had come (as much as it ever would) to Central City and its neighbors. They told a quiet story now, of what used to be, of ambitions that had been reconciled to the routines of a workaday world.

Clear Creek County also had more than held its own during the eighties, the best decade it had ever had. Silver mining led the way, complemented by steady, if not spectacular, gold production and increasing lead totals. These were the elegant eighties for George-

68. Jamestown, in Boulder County, had its moments of fame, but never emerged from mining-camp status. In the midst of prosperity, when this 1884 photograph was taken, some of the locals presented themselves in their Sunday best. Unusual was the appearance of both a black and a Chinese, who generally were not found in group pictures like this. Courtesy the Boulder Historical Society, Boulder.

69. Silver Plume shows what a fairly prosperous middle-sized camp looked like in its prime. The railroad boomed this community and the freshly painted buildings show that the prosperity has not yet worn off. Courtesy the Denver Public Library, Western History Department.

town, which could be dubbed the dowager silver queen. Some of the finest examples of Victorian architecture were to be found within this community, both in private and public buildings. Though it lost out to Leadville and other new silver districts, it retained a position as one of the principal silver areas. The coming of the railroad and the renewed interest in silver explain why this district finally achieved recognition after the years of frustration and experimentation with local silver ores.

In Clear Creek Canyon, a short distance from Georgetown, Silver Plume also secured rail connections in 1882 over the famous stretch of track that became known as the Georgetown Loop. In spite of its newfound fortune, Silver Plume remained always in the shadow of its more prosperous rival and never advanced beyond the camp stage; its architecture, for example, retained the familiar frame false-front style. On the valley floor, stretching from Idaho Springs westward toward Berthoud Pass, a string of tiny camps blossomed and briefly enjoyed prosperity. Among these was Lawson, which dated back to the 1870s and now had a reduction works, stores, school, post office, and a population of perhaps 500. A few diehards were even placer mining along the county's creek beds, sluicing as much as $25,000 in 1889.

Not all the old mining regions were so fortunate; Boulder County declined, production slumping from a high of over $700,000 to $300,000 some years. The sharp decline of the county's once "peerless silver queen," Caribou, caused its share of the troubles. Diminishing ore values and the eroding silver price, plus increased mining costs, sabotaged this silver district by mid-decade. Fortunately, steady gold production kept the local mining interests afloat. The lode mines in and around the two older districts of Ward and Gold Hill, and the new ones at Jamestown, played a large role in salvaging Boulder County's production. Ward had been opened in the 1860s, but its sulphide gold ores defied profitable reduction until the 1880s; at one time in the early 1870s only twenty percent of assay value had been recovered.

Summit County, though not quite as bad off as Boulder, followed an erratic course in the 1880s, crashing from a production level of over two million dollars to three hundred thousand, then reversing itself and rallying. The railroad helped account for the silver-lead spurt of the early years, but deposits were small and soon gave out. Another impetus to the boom was the increased interest in the Leadville area, from which miners spilled over into the Ten Mile district, Kokomo and Robinson, in the southwestern corner of Summit County. Until then placer mining had dominated, but now, in the mid 1880s, gold lode mining superseded that process. Reduced shipping rates for low-grade ore, the construction of new mills, and the revival of older ones helped sustain the conversion. One report exclaimed in 1884, "the whole county seems to have new life enthused into it."

Two other mining areas need to be mentioned in passing. The mines at Silver Cliff and Rosita continued their downward course, even after the arrival of the Denver and Rio Grande. By the end of the 1880s, this once-promising district had been reduced to a minor role in Colorado mining, its checkered career giving a clear lesson on the fluctuation of the mining industry. Park County, in contrast, produced steadily, if not spectacularly, operating both gold and silver mines.

Even placer operations continued, some being run by Chinese labor, an unusual phenomenon since the Chinese were rebuffed in many Colorado mining regions. It was not simply a racial issue, although that played a part; the Chinese were looked upon as harbingers of a district's decline. People feared that they could earn wages where a white man could not, and that they would work for less, thus undercutting wages. A disgraceful thread of violence against the Chinese runs through Colorado mining history, and as late as the early twentieth century, they were unwelcome in some communities.

This was Colorado's precious metals and lead mining in the 1880s, a decade in which it finally gained the pinnacle and became America's foremost mining state. These were years of growth, stagnation, or decline, depending on where one looked and when. A current of uneasiness manifested itself in silver mining; the nagging price decline increasingly threatened to undermine the state's principal industry. By 1889, spokesmen were bringing the issue into focus, and a silver banner was about to be raised.

Smelting continued the trend, begun earlier, toward regional centers. Leadville, Denver, Durango, and Pueblo grew, but Golden declined—it was too near bigger works in Denver. The original idea of local smelters died hard, and the noted mining engineer, Thomas A. Rickard, on a tour of the San Juans in 1902, commented on the ruins of smelting furnaces, doomed from the start because of poor locations. They were fated, he wrote, "to point a moral and adorn a melancholy tale," serving only to train "many of our best men." He waxed romantic over these early efforts which had a personal equation and a human interest "lacking in the larger undertakings which succeeded them." Neither the investor who lost money, nor the miner who lost ore, would have been so generous.

Colorado was making progress in the application of science to mining. The Leadville bonanza paved the way for Samuel Emmons's monumental monograph, *Geology and Mining Industry of Leadville,* which became known as "the miners' bible." Emmons's work,

70. In the 1880s tramways became popular to help overcome transportation problems and costs. Not all of them involved towers and cables snaking around mountains; for example, note this one at Calumet Iron near Salida. Courtesy the Kansas State Historical Society, Topeka.

Rickard concluded, taught the practical usefulness of correct geologic diagnosis and showed the significance of scientific conclusions. More recently, mining scholar Rodman Paul agreed and went further; the publication "convinced skeptical mining operators that they could learn something of cash value from university men." This in itself was a major concession and the Colorado Scientific Society, organized in 1882, was an early manifestation of this change. Improvement in smelting practices and the utilization of trained mining engineers throughout the industry evidenced a new awareness by Colorado miners and milling men.

At a less spectacular pace, both Colorado oil and coal were gaining ground. Colorado coal production passed the one-million-ton mark and reached a ten-year high of two and a half million tons in 1889, despite the fact that the price of a ton of coal on the average dropped nearly 70¢ (to $1.54), down considerably from the price of the 1860s. With the continued growth of railroads and the unending opening of precious metal mines, the statewide market had jumped as predicted, and Colorado coal was being sold in Kansas, Nebraska, and Texas.

Camps also developed around these mines. They were different from their hardrock cousins in that they quite often became either company-owned or company-controlled. Little of the free enterprise that characterized the life of the gold and silver camps appeared here. The nature of coal mining meant that the company owned large acreages of land on which were home and town. The coal miner generally did not exercise the freedom of movement nor the individualism of his hardrock counterpart, being typically recruited from a different ethnic background and handicapped by language and cultural barriers. He recognized that the next coal camp would be much the same for him, because he was simply a working miner in an industry that came under nearly complete corporate domination amazingly early.

The railroads came to dominate the coal fields, the Denver and Rio Grande leading the way. It was a natural alliance and resulted in more rapid corporative control. The railroads were especially active in opening mines along their rights-of-way and the steady increase in production reflected these new beginnings rather than improvements on the old. Most of what was mined was bituminous coal; the only developed anthracite mines were in the Crested Butte area. Las

71. Coal was becoming an important mining product. For smelting and assaying, coke was essential because of the high heat it created. This man was working on coke ovens at Crested Butte. Courtesy Frank Zugelder, Gunnison.

Animas and Huerfano counties led the way, but coal had been found in many parts of the state and, because of mounting demand for coal, the future looked more promising than ever.

The rise of coal mining brought with it a fearful increase in the number of deaths in the mines. The worst tragedy occurred at Crested Butte, where, on January 24, 1884, fifty-nine men were killed in a gas and coal dust explosion. In 1889, twenty-four men died throughout the state in coal mines, ten at the White Ash Mine (near Golden), when the workings flooded. Coal mining, by the nature of the mineral itself, with greater hazards of gas, explosions, and fire, was more dangerous than hardrock mining. The state had enacted a Coal Mines Act (1883), later amended, patterned after laws in Pennsylvania and eastern coal states. The statutes looked good only on the books.

Oil finally made a splash, to use a bad pun, late in the decade with the emergence of the Florence field as a profitable producer. The organization of companies, some with adequate financial backing, sparked a period of drilling, which uncovered a productive zone between 960 and 1,910 feet deep. Not all wells turned out so successfully. One dry hole reached a depth of 3,100 feet, deep by the standards of that time, and another was a poor producer after costing $20,000. Outside expertise and machinery had been sought; for instance, a complete drilling rig was purchased in Pennsylvania. In 1887 a local refinery was built and in 1889 it had been joined by another, when the field's yearly production reached 316,476 barrels. By Colorado standards it was a boom, even if eastern oil men looked upon Florence as a poor country cousin of the nation's oil business.

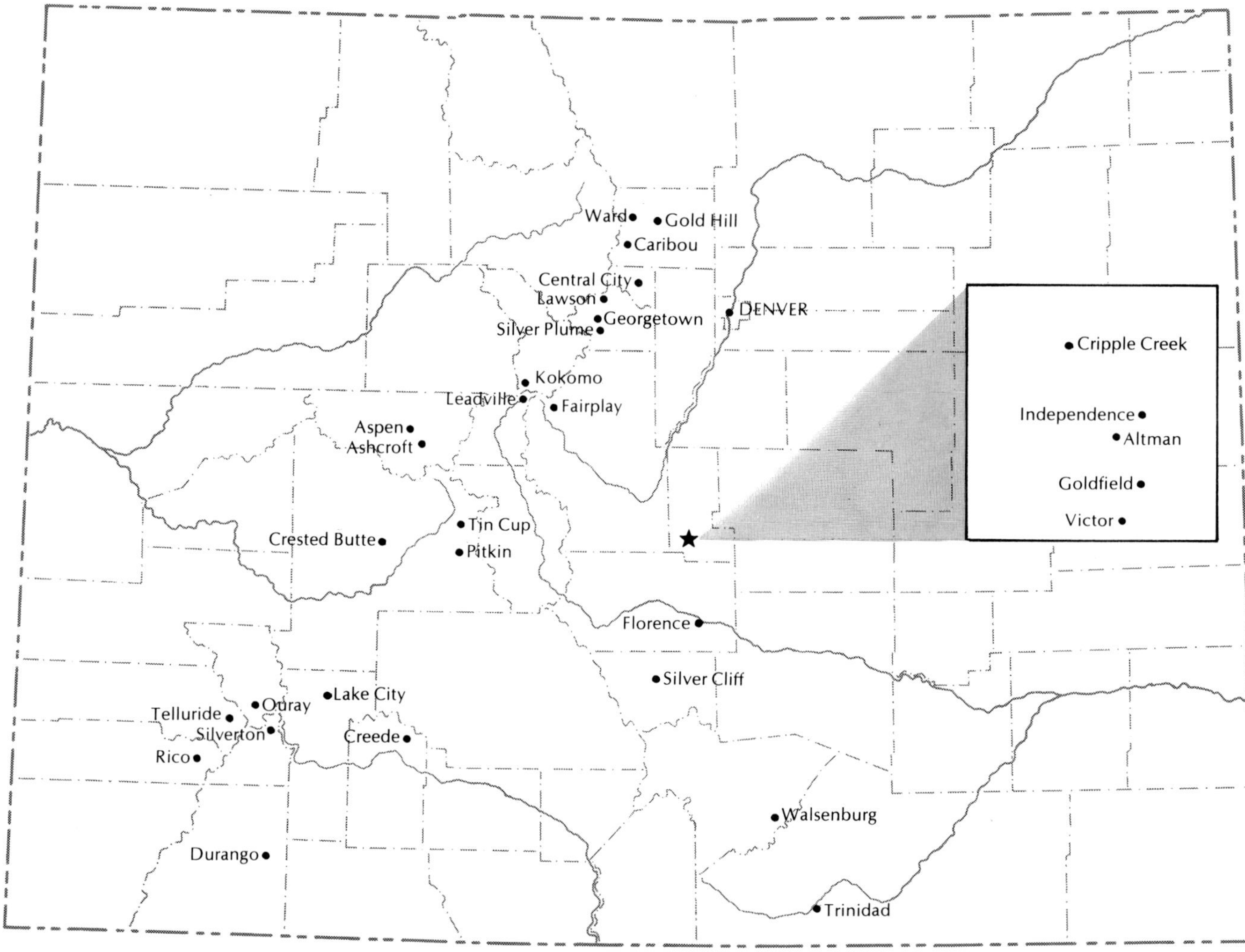

MAP 4 Colorado Mining, 1880s–1890s

6

Gold to the Rescue

Gold was a magic word in the 1890s, ringing throughout Colorado and spreading through the country and beyond to eager listeners in England and Europe. A generation had sought it since 1858–59. Where once profits of thousands of dollars had tempted investors, now hundreds of thousands, even millions, lured them. Thus it was that Cripple Creek became to Colorado in the nineties what Leadville had been in the seventies; gold and Cripple Creek were indistinguishable. Cripple Creek, the state's greatest gold camp, proved a worthy successor to declining Leadville; hardrock mining predominated—the pan and shovel had little place. This, the last hurrah of the rapidly fading mining frontier, enveloped the state at a crucial time.

For two decades silver had dominated Colorado mining. Twenty years was a long time for a mine or a district to maintain peak production, and many besides Leadville were showing their age. Enough new mines were being opened, however, to keep production high, mounting in the early 1890s to $20 million per year, the highest total in Colorado's history. Aspen sparkled in all its glory, replacing Leadville as the new silver queen, although it was forced to share some attention with Creede, in recently created Mineral County on the eastern border of the San Juans. Such old producers as San Miguel, Ouray, and Clear Creek counties mined most of the rest of the silver. Had the price of silver remained at its 1870 levels, only decreasing ore reserves would have threatened local prosperity. But government support failed to maintain the price, and it sank steadily from the $1 per ounce range to as low as 59¢ average in 1898. The steepest fall came in 1893–94, along with a national crash and depression. This devastating combination hit the silver camps and mines severely, closing the smaller and older mines, throwing men out of work, depressing local economies, and generally casting a pall, which had not been duplicated since the dark days of the early and mid-1860s. For the first time in over twenty years, no new silver districts were being opened, no booming district grabbed the attention of the miner, merchant, or laborer. At this point Cripple Creek sallied forth, distracting, for a while, Coloradans' attention from some of the serious problems besetting their mining industry.

Cripple Creek was situated just southwest of Pike's Peak, in an area generally ignored since the days of 1859. In 1884 there had been some furor over purported gold discoveries, but the early birds who rushed in discovered that the claims had been salted, apparently to stimulate business in Canon City, which advertised itself as the gateway to the new mines. The Mount Pisgah hoax (as it became known) gave the region a bad name and must assume a large share of

72. Times had changed and a visitor could ride in comfort to the site of a mine. So could reporters and Creede gained instant fame. Courtesy Western Historical Collections, University of Colorado, Boulder.

73. The pace of Colorado mining was varied. Districts were booming, stabilizing, or declining, depending on where one looked. Central City matured with age but no longer was the number one gold district. Courtesy the Denver Public Library, Western History Department.

74. Cripple Creek was the big news in Colorado in the nineties. At first, placer mining—reminiscent of 1859—was tried. Only limited success awaited these optimistic people; the real money was in hardrock mining. Courtesy First Federal Savings, Denver.

75. Millions poured out of the Portland into the pockets of Winfield Stratton, James Burns, and others. Mining at Cripple Creek was big business from the start, and there were plenty of vultures ready to pick the carcass of an unwary owner who happened to strike it rich. Thanks to Stratton, the Portland survived the attacks. Courtesy the Colorado College Library, Colorado Room, Colorado Springs.

the blame for delaying interest in later discoveries. Cattle and cowboys roamed this land which seemed destined to remain a cattle range. A few prospectors came and went and one, Bob Womack, also a part-time cowboy, found promising gold-bearing float. Unfortunately, he was an undistinguished individual given to carousing, and little credence was attached to his statements. Finally, in 1890, after more than ten years of searching, Womack opened a claim yielding ore which assayed at $250 per ton. That was something substantial—more than a mere curiosity—and interest was aroused on the other side of the mountain at Colorado Springs.

More people came, and by the summer of 1891 claims were staked on nearly all the hills in the immediate vicinity of Womack's activity. In April, a mining district had been organized, named after the geographic feature most closely associated with the land, Cripple Creek. With both United States and Colorado examples to copy, mining laws were quickly formulated. Quite a few tenderfoots were among those heading towards Cripple Creek, but even prior experience with earlier discoveries and prospecting was no guarantee of success. The gold-bearing lodes ran so contrary to previous Colorado occurrences that the amateur was likely to do as well as the expert in the game to find the paying claim. Located in a ten-thousand-acre bowl of volcanic rock which had been fragmented by eruptive explosions, Cripple Creek showed little surface gold. The ore veins were relatively narrow and not marked by quartz outcroppings, nor sufficiently different from surrounding rock to call attention to themselves. Only by stripping off soil and loose rock, or sinking test shafts, could veins be discovered. Valuable ore was sometimes thrown out on the dump and, conversely, worthless gangue was shipped in error. Prospecting required continual assaying, a costly procedure for a man such as Womack; the district's large number of stamp and sampling mills attested to the fact that the same problem troubled large companies.

The mere finding and filing of claims made no one wealthy—it took capital to open and develop these properties. Cripple Creek was no "poor man's diggings." By the fall of 1891, with money coming in, Cripple Creek was on its way. It had now become more than just a district—a camp had sprouted out in the flats near the spot where Womack had made his discovery. Soon other little communities joined it, sitting like nails in a horseshoe-shaped curve bending southeastward to Victor and then back toward Cripple Creek. This area of approximately twelve square miles became one of the most urbanized in Colorado.

The primitiveness which characterized the earliest days of Colorado mining communities passed very swiftly here and at Creede. Horace Greeley would have been incredulous, as were most visitors, at the rapid changes. The railroads came quickly; in fact, they brought the tourist to see the sights almost before the sights became seeable. Cripple Creek, once its worth was demonstrated, became almost irresistible. Its population reached a reported 5,000 by 1893, and the whole district topped 20,000, according to the census takers of 1900. Victor, situated nearer the major mines, never was able to match its rival in either reputation, population, or wealth, and was always more the miners' community, without the frills which attracted more management and wealth to Cripple Creek. The other smaller communities lay in the economic orbit of one of these two.

Cripple Creek was no hoax. Production there was valued at $500,000 in 1892; in 1893 it shot to over $2 million; and in 1899 it reached a decade high of $16 million. Cripple Creek produced more gold than had ever been mined before in Colorado and was a worthy rival to the California Mother Lode country. It was ironic that such wealth was found within the shadow of the peak that gave its name to the 1859 rush, and which had promised, but never delivered, this kind of success.

Among the discoverers of bonanza mines was Winfield Scott Stratton, long an unsuccessful carpenter-prospector, who had toured many of Colorado's mining districts before coming to Cripple Creek. There he started the Independence Mine, which made him a millionaire several times over before he sold it in 1899 for $10 million to English investors. Stratton headed a company of at least twenty-seven, all of whom reached millionaire status, thanks to Cripple Creek gold. His Independence Mine, rich as it was (it produced over $3 million by decade's end), was still second to the Portland, which topped eight million in the same period. Shrewd and cautious, Stratton also owned the largest share of stock in this latter property, and he became, in his own way, what Horace Tabor had been earlier.

Even Cripple Creek was not without rivals for investors' capital. A concerned newspaper editor wrote in the December 16, 1897, edition of the Cripple Creek *Morning Times* that the East was being flooded with pamphlets and literature about the Klondike gold discoveries. He warned that companies were being organized to divert money to this "unknown arctic region," and that they would never return a penny. "Why not divert as much of it as possible to Cripple Creek, where results are certain," he reasoned. "It will only require the magic touch of money to start dozens of mines here on a career of production that will astonish the world." The Klondike bloomed and faded before another summer was spent, but the concern expressed had been typical of many Colorado districts. Editors, promoters, and miners had always been jealous of money and investors that looked anywhere besides Colorado.

Cripple Creek had not peaked; 1900 would be the best year. The last and greatest of Colorado's nineteenth-century gold rushes ushered the old century out with a final flourish. Three major pinnacles of that century stand out: the initial excitement (centered in Gilpin County), the Leadville stampede, and the Cripple Creek bonanza. In between, others created ripples and a few waves, but none was able to grab and hold attention as these three and none proved as significant. As Leadville had turned Coloradans' attention to silver, Cripple Creek returned it to gold, where it stayed until the First World War.

The attention of Coloradans needed to be turned to something other than the economic and political problems that confronted them during the nineties. The crash of 1893 staggered Colorado, severely damaging silver mining in the mountains and hurting agriculture on the plains. Business failures, bank closings, and mortgage foreclosures were the obvious signs of the times; nearly everyone in the state was affected as Colorado was one of the hardest hit in the country. Even without the depression, many of the older silver districts would have faced hard times simply because of their age and a declining international silver market. It was unfair to blame the 1893 repeal of the Sherman Silver Act, which provided a government-guaranteed market, for killing Colorado silver mining. A quick check of statistics reveals that production of silver was higher in the 1890s than ever before or since. Those mines in high-grade ore continued to produce, and silver was increasingly mined as a by-product of gold, especially in the San Juans. The marginal, older, and smaller mines were undercut and their closure hastened by repeal.

For Coloradans and others it was easier to emotionalize hard times rather than to analyze their causes. It was much easier to blame the silver problems on an eastern Wall Street conspiracy, an international financial plot, or the business-oriented Republican party in control in Washington, than to sit down and examine the situation thoughtfully. Silver had surfaced suddenly as a miracle cure-all for the hard-pressed miner and the debtor farmer, both of whom extolled its virtues rhetorically rather than logically. Silver and agrarian spokesmen and women rallied their listeners around a silver banner, raised it high politically, and saw it ripped down when silver and its presidential candidate, William Jennings Bryan, marched to defeat in the hotly contested 1896 election. Never again nationally would a candidate's stand for placing silver on parity with gold and raising the price of silver be the touchstone of his success or failure. Colorado's silver years came to an end during the furor surrounding the silver question. Silver had lasted a strong twenty years and carried the state far, but it had lost its fascination for the public. In mining circles, interest persisted and the search for and mining of it continued, but at a gradually slackening pace.

The new interest in gold mining produced fascinating results in some of the state's older mining regions. For instance, Lake County's gold production jumped immediately in 1893 and reached the $2 million level by 1897, up ten times over what it had been five years before. The gold properties were located chiefly on Breece Hill, and talk of a district "gold belt" aroused fresh hopes before reality quieted them. A small, growing interest in zinc augured well for the future, as did the fact that the Leadville mines kept active throughout the trials of the nineties. Leasing and

76. In the 1890s for the first time labor violence seriously hit Colorado's hardrock mining. The Strong Mine was blown up in May 1894 to convince some professional toughs that the strikers meant business. The Cripple Creek District would be divided into two embittered camps by a decade of violence that was just beginning. Courtesy the Cripple Creek District Museum, Inc., Cripple Creek.

consolidation of older properties were features of the last years of the century in a district which one 1899 observer believed was "never more active since the famous carbonate discoveries of 1878."

Even more spectacular was what occurred in the San Juans, led by San Miguel and Ouray counties, followed closely by San Juan. Ouray's impetus came from the Camp Bird, one of Colorado's famous mines, which was opened in 1896 and produced $2 million by 1900, mostly in gold, but with a smattering of silver, lead, and copper. The Camp Bird bestowed on its owner, Thomas Walsh, enormous profits, stated to be sixty-five percent of the value of bullion and concentrates. He reaped six million dollars when the property was sold in 1902. Charles Henderson, in his famous study *Mining in Colorado,* wrote that the Camp Bird profit "probably represented the highest percentage of gross value of output of any mine in the state."

Across a twelve-thousand-foot ridge southwest of the Camp Bird were San Miguel's famous threesome: the Tomboy, Smuggler Union, and Liberty Bell. All of these had been discovered in the 1870s, but profitable working proved impossible until transportation improved and adequate financial support was secured. With the coming of the Rio Grande Southern Railroad to Telluride, and the influx of outside capital and purchasers, these three reached their potential in the late 1890s. They carried with them San Miguel County, which after 1897 was one of Colorado's major producing counties, gold and silver being the principal minerals, followed by lead, zinc, and copper.

The San Juans, thanks to Creede and the emergence of its noted gold producers, finally received the adulation they had long courted. Even Cripple Creek, in all its glory, could not completely outshine them, and Silverton, Ouray, and Telluride were three mining towns which prospered throughout much of the troubled nineties. Not all the San Juans were as prosperous; the Red Mountain mines declined, as did those in Rico and most of the smaller camps scattered throughout these mountains. Gilpin meanwhile had a solid, if not spectacular decade, the last before it began to wane. This county remained a prime gold producer for an extraordinary length of time. Only the emergence of Cripple Creek pushed it from its number one position in Colorado, and eventually completely overshadowed it in total production.

Also eclipsed at the moment was the introduction of dredging in 1898–99 in the Blue and Swan rivers. Dredging seemed the ultimate answer to profitable working of the low-grade placer areas. A power-driven chain of small buckets mounted on a barge or boat, anchored in its own pond in the stream, could work large amounts of gravel. A washing or treatment plant was also on the boat, which crept and cranked its way along the stream, leaving in its wake squalid piles of washed rock to mark its course for decades to come. The dredges overcame the problems of boulders, rocks, and large acreages of low-grade gravel and became landmarks of the Breckenridge and Fairplay areas. The first ones proved too small and light, and modifications had to be made. Successful and profitable dredging operations awaited the twentieth century.

Colorado had had a remarkable record of labor peace in the years since the original rush. The Leadville outbreak of 1880 had been the only one of major proportions in the hardrock industry, although smaller ones had occasionally flared in response to local working conditions and wage scales. Coal mining, with a national record of bitter strikes, had been much more volatile. Such issues as wages, unfair firings, and demands for miner-appointed check weighmen to keep tabs on the company scale operators (coal miners were paid by tonnage mined) caused the miners to walk out. Five times in 1884 and six in 1886, for example, they struck, supported by their union (usually only a local one, although some men affiliated with the Knights of Labor and a few with the Western Federation of Miners later). Again in 1893–94 there was a rash of strikes in the various coal fields, this time against company stores and company-issued scrip redeemable only at those establishments.

The growth of corporative control and absentee ownership, accompanied by the decline of individual opportunity, bespoke a changing situation in the hardrock mines as the 1880s passed. Men were no longer miners in the sense they once had been. They had become laborers in nineteenth-century industrial America, employed by someone else for wages prescribed by management.

The opportunity to move to a new district eroded steadily; few new ones, except Cripple Creek, persisted into the 1890s. A dissatisfied man, who could once have moved on to Central City, to Leadville, to Telluride and beyond, if working conditions or wages did not meet his expectations, now found this escape route closed. Indeed, as miners swarmed out of declining silver districts, Cripple Creek was the only Mecca left; even the prospering older areas could not absorb more than a fraction of those looking for work. In 1892–93 a knell sounded, as strikes hit Aspen, Rico, and Creede over hours, wages, and unionization.

With a growing labor surplus in their favor, the Cripple Creek mine owners moved to reduce wages from the standard \$3 per day to \$2.50 or, alternatively, to increase the work day from eight to ten hours at the old wages. Living costs were high (though not up to pre-railroad heights), and a man could barely maintain his family on \$3, provided he worked steadily, which many did not. In 1893 the threat of a ten-hour day promptly forced the miners to organize a local union and affiliate with the Western Federation of Miners.

This union gave the Colorado hardrock miner, for the first time, a strong voice, dedicated leadership, a platform, and the possible support of fellow union members throughout the Rocky Mountains. A new element had been added to the growing feud between management and labor; now there were the haves and the have nots, ready to stand forcefully behind their positions. Unable to agree on districtwide wages and hours, a group of owners took the offensive in January 1894, and unilaterally instituted the nine-hour day. When neither the Portland nor the Independence followed suit, their position was plain for all to see. Union leaders, unimpressed by pious statements about hours in other Colorado mines and how pleasant it was to work in Cripple Creek's dry, well-ventilated works, struck the mines. They stayed out in the face of threats and pressure, defying the owners from their headquarters at Altman, the most unionized camp in the district. Finally, the deteriorating situation forced Governor Davis Waite to send in the state militia; he withdrew them after apparently calming both sides. But the peace was only transitory. When the surface workings of the Strong Mine were blown up, participants on both sides roughed up, stores robbed, and little armies organized, Waite sent the guard back. The governor, whose political views had already made him unpopular with the owners, aroused tempers by siding with the miners. Finally, after more than four months, an agreement was reached with the miners that gave them an eight-hour day at \$3. The Western Federation had won a strike, one of the longest and fiercest American labor disputes up to that time. The owners learned lessons they would not forget, and the union gained new prestige. The cost in lost wages and production, state expenses, and bitterness had been high and the contest was not over; the war had just begun.

The struggle shifted to Leadville where, during the depressed times of 1893, the miners had agreed to \$2.50 per day as long as silver was bringing less than 83½¢ per ounce. Only those desperate times forced acceptance of what was considered a poor wage. By 1896, some gold-silver mine owners had abandoned that scale and were paying \$3 to obtain skilled miners. Caught in a profit-expense bind, they hinted that year that, if all operators did not come to that level, they would have to reduce theirs. Flushed with victory, the Western Federation had organized this town. Riding a crest of popularity, it checked this threat by calling a strike for a Leadville \$3 wage. Both sides dug in, the mine owners secretly organizing an association, fortifying their properties, and bringing in strike-breakers or scabs, while refusing to recognize or deal with the Federation. The situation worsened rapidly, both above and below ground. When work stopped, the pumps were silenced and the mines flooded, retarding operations throughout the remainder of the nineties. Above ground, violence enveloped several mines operated by non-union workers; in September a mob attacked the Coronado and Emmet, costing several lives. The call went out for help, and this time the governor, Albert McIntire, sided with management. In came the national guard troops, whose presence allowed the mines to be gradually reopened. The strike, meanwhile, dragged on to a tortuous end, the miners finally conceding and dejectedly going back

77. Once the owners realized the importance of having a favorably inclined state government in office, the labor struggle became an unequal contest. Colorado National Guard camps were familiar sights, as the troops were sent in to calm things down, and they worked almost exclusively on behalf of management. These troops garrisoned Leadville in 1896. Courtesy the Lake County Civic Center Association, Leadville.

78. Fire, one of the dreaded scourges of the hastily built wooden camps, could mean the end for a declining community. Conversely, it might produce a better arranged and constructed camp if the district continued to prosper. This photograph shows Victor during the fire of August 21, 1899. Courtesy the Cripple Creek District Museum, Inc., Cripple Creek.

to work on the employers' terms. The union was broken. Owners had learned that by organizing themselves and securing a friendly state government the game could be tipped in their favor; they did not forget that lesson. The miners were left with a bitter aftertaste: hatred came to replace mere dislike, and violence superseded peaceful solution.

In 1899 a small strike over company rules in Lake City led to trouble with recently hired Italian miners. It ended quickly in a company victory. Noteworthy, as a sidelight, was the fact that the national character of a large segment of the Colorado mining population, both hardrock and coal, was becoming southeastern European. This change was more striking in hardrock mining where English, Irish, northern Europeans, and native Americans had once strongly predominated.

The smelters also employed large numbers of Italians, Austrians, Serbs, and other nationalities, as both skilled and unskilled laborers, giving a new flavor to the settlements huddled around the plants. Smelting had become increasingly more complicated with the introduction of new processes that involved chlorination, cyanide, and a battery of new machinery. The smelters and mills needed to handle the Cripple Creek-Telluride gold ores created a booming industry for Colorado Springs, Florence, and Canon City. Smelting had become big business, not only in Colorado, but throughout the United States as well. As such, it attracted the attention of skilled, profit-seeking businessmen who were determined to gain control of it, much as John D. Rockefeller had done with oil, and as other American entrepreneurs were doing on a smaller scale elsewhere. The 1890s were a good time for consolidation; solvent corporations were easily able to buy out less successful rivals in an effort to achieve larger and more efficient organization. Trusts emerged as an emotional political issue and a fact of life in American business of the 1890s. More crucial to

79. Victor after the August 21, 1899, fire. Courtesy the Cripple Creek District Museum, Inc., Cripple Creek.

80. Victor ten days after the fire; the town started to rebuild immediately. Courtesy the Cripple Creek District Museum, Inc., Cripple Creek.

81. By the 1890s, smelting was centralized in a few important cities, Denver being the principal one. Hill's Argo Works, pictured here, was still one of the best. Around the smelters were the homes of the workers, often recently arrived immigrants from Eastern Europe. Courtesy the Denver Public Library, Western History Department.

Coloradans, however, was the silver issue; no trusts seemed to threaten them as much as what might happen to silver.

Then in April 1899, Colorado suddenly awoke to a realization of what it meant to confront big business and monopolistic control. The American Smelting and Refining Company, incorporated in New Jersey (as were numerous mining companies in that state of few regulations), acquired many of the largest smelting plants across the entire country. Included were the Denver, Pueblo, and Durango smelters; overnight, this company became Colorado's single most important smelter corporation. Not all the smelters had been absorbed, but enough to cause dismay among miners, mine owners, smelter workers, and consumers generally. No union or mine was strong enough to challenge this conglomerate, except perhaps in Cripple Creek. A state the size of Colorado could hardly attempt to regulate such a giant. Coloradans became justifiably concerned as the decade and the century ended.

Though Colorado might have had difficulty regulating the American Smelting and Refining Company, it was doing better in the field of mine regulation. In March 1895, the legislature created a state Bureau of Mines and Commissioner of Mines to replace an "inspector of metaliferous mines," a weak office which had been set up six years before. The new commissioner was charged with supervision of mine inspection and enforcement of mine safety and health laws. To keep him from being idle, his jurisdiction was broadened to include mills, smelters, sampling works, rock quarries, and railroad tunnels—almost everything except coal mines. The Coal Mine Inspector took care of that aspect of the industry, as he had done since 1883. The Bureau was also designated to collect and exhibit mineral specimens, mining data, and books—projects which were often shunted aside under the pressure to accomplish main objectives.

In the field of individual mine and district reports, the high quality tradition of Samuel Emmons was continued by a dedicated group of men, including at the state level, Arthur Lakes and Harry Lee, the first commissioner of mines, and, for the United States Geological Survey, Whitman Cross and Chester Purington. Colorado was fortunate to attract such scholarly and professional interest, before the older mines were completely shut down and the districts reduced to mere shadows of their former selves.

82. Florence was beginning to look like an oil boom town with wells located in the community's heart. The Florence Field was at its peak. Courtesy the Denver Public Library, Western History Department.

The transitional 1890s ended as they had begun, on an upturn. Gold, rather than silver, now provided the thrust. The *Weekly Denver Republican,* on January 4, 1900, summarized the year just past with bold headlines, "Colorado Mineral Output Greatest In Its History." The story concluded that "the mining interests of the entire state have never been in better condition."

Cripple Creek's great rush ended an era that went back to 1859, while the strikes and smelter trust struck discordant notes that would echo far into the century ahead. Not since the 1860s had such troubled times threatened Colorado; not since the 1870s had new discoveries precipitated such exultation. It was, however, the last stand for much that was being discarded in the rush toward an industrialized America. Cripple Creek furnished possibly the best opportunity to sense the undercurrents of what was to come, and of what had been. Leadville now looked less like a mining community and more like a typical industrial city of the steel-mill region of Pennsylvania. It was hard to see how the tired, dirty miner working his shift had more freedom and individualism than his counterpart in the steel mill. Management, looking on from its panel-lined conference rooms, treated them all much the same.

7

The Business Called Mining– A Photographic Essay

That mining included more than just digging the ore out of the mountainside, panning in a stream bed, or drilling in the earth should be obvious. Mining was never a simple industry and rapidly became more complex after the days of 1859. It evolved into a modern industrial corporation, one which rises and falls partially on the success or failure of the miner and partially on national and international factors. The "pan handling" man of earlier days would not easily recognize his offspring.

Even in the nineteenth century, however, evidence was plentiful that mining involved many factors. No Colorado hardrock district could flourish without a railroad or a generous injection of outside capital. Three ingredients were vital—dependable year-round transportation, financial support, and profitable ore. Only then could needed mills and smelters be constructed, mines developed, the district advertised, and settlement put on a semipermanent basis.

As the mines opened, the need for lumber increased, stimulating that industry. As the shafts went deeper, new machines, pumps and hoists were required, which added another business to the local or state economy. Mining was the heart of an ever widening business that affected the railroader, the farmer, the Denver businessman, and the New York and English investor. The miner laboring in the mountain's heart was not working simply for himself. Whether he realized it or not, many hopes rode on the ore car as it carried his shift's production on the first leg of its journey to becoming bullion. Correspondingly, he could not mine without the help of outsiders; they were all tied together, related by misery and joy, in the business called mining.

83. The prospector and his burro are parts of the legendary element of Colorado mining. These men opened many districts and found mines, only to sell out and drift on, caught in the wanderlust of mining. Their burros, according to legend, helped them find mines through stubbornness. More accurately, these beasts proved to be hardworking, ill-appreciated pack animals. Courtesy the State Historical Society of Colorado, Denver.

84. In the days before faceless corporations took over mining and smelting, men were known for their wealth and success. Quite often they emerged as political, business, and social leaders as well. Their day of fame spanned the nineteenth and early twentieth centuries when Colorado mining was new and prospering. Legends grew up around some of them, and they were the best known Coloradans of their generation. Thomas Walsh was one of these men. After years of limited success, he made his fortune at the Camp Bird Mine and found fame in Colorado and Washington. Courtesy Ruth Gregory, Ouray.

85. Albert E. Reynolds owned mines in the San Juans, Gunnison country, and elsewhere. Not so well-known as the others of his generation, he represents those whom notoriety and wealth touched less abundantly. From the author's private collection.

86. Simon Guggenheim, a member of the smelting family, worked in Pueblo and Denver. Wealth and political achievement came to him. From the author's private collection.

87. Winfield S. Stratton's Cripple Creek mines made him the richest of all the mining kings. From the author's private collection.

88. Nathaniel Hill got his start in smelting, but he also dabbled in mining, politics, and newspapers. He was the father of scientific smelting in Colorado. Courtesy the State Historical Society of Colorado, Denver.

89. Horace Tabor's career epitomizes the Colorado mining frontier. A legend in his own time, he remains one today. Courtesy the Library of Congress, Washington.

90. The Tabors and Walshes worked in the mines, but once their wealth came, they turned the operations over to managers. These are the mine bosses at the Cresson Mine in Cripple Creek, and they appear quite stern. Courtesy the Cripple Creek District Museum, Inc., Cripple Creek.

91. At first, trained mining engineers were mistrusted. Practical experience and common sense were deemed more important than academic training. As the years passed, engineers came to be accepted and with them came the agreement that professional education and training could be of value. These two, Carl Foster and Everett Shelton, worked in Taylor Park in the Gunnison country in 1908. Courtesy Carl Foster, Ashby, Mass.

92. Thomas Walsh spoke from experience when he said: "The advent of the mining engineer has lifted mining from a plane of reckless speculation to that of legitimate, safe, and lucrative investment." This is a group of Columbia School of Mines students on a field trip to Colorado in 1895. Courtesy the Columbiana Collection, Columbia University, New York.

93. George Purbeck called himself a mining engineer, but he was more a promoter and mine speculator. He tried to sell the Caribou Mine in the 1890s, without much success. A mine speculator was defined as a "man who worked claims with his jaw instead of with his pick." Courtesy the Boulder Historical Society, Boulder.

94. The owners, managers, and engineers would have planned in vain without the sweat, brawn, and skill of the men who labored eight to twelve hours a day in the depths of the mines for a wage that often just barely provided subsistence. On their efforts rode the hopes of the stockholders who knew little of the dangers or work involved. No workman's compensation or unemployment insurance eased their worries, and mining safety laws appeared only slowly; enforcement was another matter. The miners of the Louisville Mine near Leadville, before going to work on May 25, 1888, are shown here. Note the lack of hard hats, the equipment, and the uniformity of clothing. The boys worked on the surface although the very youngest one probably just posed with his dad. Courtesy the Lake County Civic Center Association, Leadville.

95. This is the day shift of the Republic Mine in the Cripple Creek District, with the foreman or manager in front. The little fellow with the cigar was pulling someone's leg. Courtesy the Cripple Creek District Museum, Inc., Cripple Creek.

96. Chinese were often driven out of Colorado camps because of racial prejudices and because they were feared as harbingers of decline. This crew worked at a mine near Idaho Springs: others were active in the placer areas around Fairplay. Courtesy the Denver Public Library, Western History Department.

97. The underground crew depended on the surface crew to keep them mining. The blacksmith sharpened bits and drills; the hoist engineer raised and lowered men, ore, and equipment; the carpenter provided a variety of services; and the common laborer shoveled, carried, and pushed a number of items. This is the Ute-Ule crew outside Lake City. Courtesy the Denver Public Library, Western History Department.

98. Later the blacksmith shop became the tool-repair shop, with new equipment but old functions. Courtesy the State Historical Society of Colorado, Denver.

99. If an army marched on its stomach, the miner labored on his. The cooks were vital to maintaining goodwill and a solid measure of contentment. These Smuggler Union Mine cooks and helpers do not look as though they would take many complaints from patrons. At the isolated mines, boarding houses were home to most of the single miners. Room and board were deducted from monthly wages or were part of the wage package. Courtesy Mrs. Homer Reid, Telluride.

100. Management held firm control of Colorado mining, leaving miners without much of a voice until the rise of the unions. The decades of violence, from 1894 to 1914, reflected the struggle between the two. This group of striking coal miners in Boulder County had been sentenced to a year in jail for contempt of court. Courtesy the Denver Public Library, Western History Department.

101. The coal strikes were in many ways the bitterest. "Mother Jones", an eighty-three-year-old labor agitator, led this Trinidad parade in 1913 on behalf of the striking coal miners. The Ludlow tragedy eventually grew out of this conflict. Courtesy the State Historical Society of Colorado, Denver.

102. Unions survived the pressure of management and public opinion to represent most miners by the mid twentieth century. The Gilman/Minturn union local 581 held a meeting in the 1950s: it must have been a social affair, because women miners did not appear until the 1970s. Courtesy the Western Historical Collections, University of Colorado, Boulder.

103. "Promote, Advertise, or Die" could have been an appropriate motto for the mining districts. The owners realized that they had to attract buyers, investors, and outside money, and Coloradans were not reticent about advertising. Newspapers promoted, articles appeared, and speakers orated. Samples of rich ore proved particularly irresistible, as this sample from the Eagle mining district showed. Courtesy the State Historical Society of Colorado, Denver.

104. Colorado mineral exhibits appeared throughout the country and in Europe. The homefront was not overlooked: one of the major efforts produced the National Mining and Industrial Exposition in Denver, 1882–84. The main building, shown here, housed exhibits from nearly all state mining districts. Courtesy the Denver Public Library, Western History Department.

105. The mine broker's office, calm in this picture, often was the scene of harried conferences in which a mining company's fate hung in the balance. The men who speculated in stocks might never go near the mines; to them it was simply a way to make money, augmented by the thrill of gambling. Courtesy the State Historical Society of Colorado, Denver.

106. The pack mule replaced the wagon as the means of transporting goods to the more isolated camps and mines. Ironton's main street was crowded on this particular morning and even a bicycler managed to get into the act. Courtesy the State Historical Society of Colorado, Denver.

107. These burros are waiting patiently for their load of about 250 pounds to be removed. When trails were poor, the jack could haul supplies in and ore out of the mines. These burros came from Creede's Amethyst Mine. Courtesy the Western Historical Collections, University of Colorado, Boulder.

108. Occasionally, when the burden proved too heavy, the horses pushed and pulled wagons. This was how cable could be freighted. Courtesy the Western Colorado Power Collection, Center of Southwest Studies, Durango.

109. The ore wagon carried the mine's product to the mill, unless a tramway provided a better method or the trail was fit only for mules. These teams operated in Boulder County. Courtesy the University of Colorado Museum, Boulder.

110. In the twentieth century the truck began to replace the wagon and burro. Delivery trucks of the Colorado Supply Company, the company store in CF&I towns, were some of the earliest. Courtesy the State Historical Society of Colorado, Denver.

111. By the 1920s and 1930s trucks were bigger and more powerful, forerunners of today's giants. While they could go many places, they needed better roads and bridges than did the sure-footed burros. This Coleman was parked in Silverton. Courtesy the Western Colorado Power Collection, Center of Southwest Studies, Durango.

112. A few companies had steam-driven tramways. This one operated in Routt County and hauled coal to Oak Creek. Courtesy the State Historical Society of Colorado, Denver.

113. Stagecoaches were not always fancy Concord stages. Almost anything was pressed into service, including sleighs. Courtesy the Boulder Historical Society, Boulder.

114. Unusual, even rare, was the mining camp interurban, or trolley car. Cripple Creek had one, and so did the coal camps around Trinidad. This Starkville Interurban stopped on its way to the camp of the same name outside Trinidad. Courtesy the State Historical Society of Colorado, Denver.

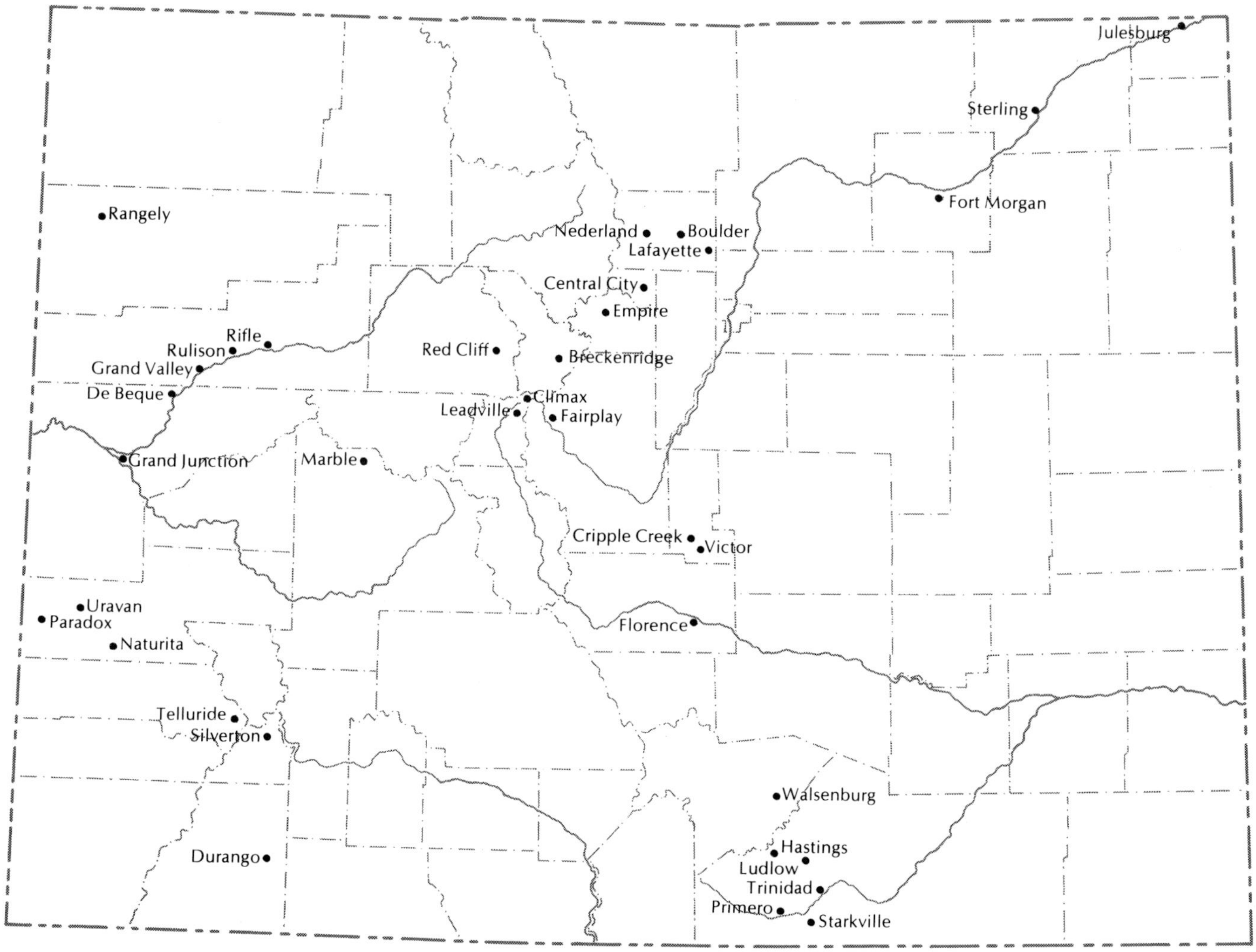

MAP 5 Colorado Mining, 1900s

8

"Damn the Owners"

With justification, Colorado mining seemed to be hurtling toward better days than the industry had known in the decade just passed. Cripple Creek, in bonanza, stayed in bonanza. In the first fifteen years of the twentieth century this district produced over $198 million worth of gold, an astonishing figure for Colorado, never before or since equaled by one district.

Cripple Creek did not stand alone. The San Juans reached new heights, fulfilling at last the forecasts of nearly three decades before. This vast district did not prosper evenly. The mountains between Ouray and Telluride were the richest. The Tomboy and Camp Bird became famous, and not far behind were the Smuggler Union, the Liberty Bell, and several other properties. When English investors came, the first two mines were quickly sold to these usually conservative people who were caught up in the excitement of gold fever. Mining still held out the promise of quick, easy wealth.

These, then, were the primary precious metals mining districts, joined by Lake County whose total was greatly augmented by zinc and lead production. Declining Aspen, and Clear Creek and Gilpin counties, trailed badly.

Mining itself was not dead, but the mining frontier faded quietly away. A certain feeling—confidence, perhaps faith—that permeated the earlier years vanished. The emergence of corporation mining and the decline of prospecting and individual opportunity undoubtedly contributed to the changed atmosphere. It had become so bad that newspapers were trying to encourage men to prospect, something they had not had to do in past decades. The "boys" of '59 were gone and the "boys" of '79 had grown old. The geographic mining frontier had long since disappeared during the 1859–60 rush; the frontier that remained was not a place, it was a time and a way of life. It had endured in those pioneers who were departing, leaving behind a rich heritage, taking with them something intangible that would be lost forever. The people who came to take their places were strangers, only vicariously able to experience what they could not relive. Mining would never be the same again.

In yet another way it would never be the same again. The years from 1900 through 1914 were ones of repeated labor violence in both the hardrock and coal districts. Corporation control and absentee ownership brought Colorado face to face with management-labor conflicts that rocked much of the rest of the country. The simmering labor unrest of the 1890s in the hardrock districts boiled over in the first years of the twentieth century. All the factors of the nineties were sharply highlighted: the increased corporation/absentee ownership, the disappearance of individual opportunity, the end of the mining frontier, and the dangerous, long working day at low wages. The mounting numbers and radicalism of the Western Federation of Miners gave a more belligerent impetus to the heated atmosphere. The owners' resentment and hatred of this organization knew no bounds. The union wanted recognition first, after which details of hours, wages, and working conditions could be hammered out. The owners stubbornly determined that they would not concede recognition.

If the miners and the union could not persuade management to agree to their demands, their ultimate weapon was the strike. What then evolved was a state of siege, each side determined to win by hurting the other economically. The owners fought back, looking upon strike breakers (or nonunion workers) as being a legitimate weapon to counter strikes. Obviously, neither side agreed with the other's tactics and violence resulted.

Telluride harvested the first fruits of this struggle. A strike broke out in May 1901, against the Smuggler Union over wages, with the Western Federation representing the miners. The company refused union demands, then promptly hired strike breakers at the terms denied the union, vividly demonstrating where the real issue lay. In July a mob of armed strikers attacked the night shift as it came off work, killing and wounding several and driving the rest of the "scabs" from the district. A brief compromise followed, only to be broken when the manager of the mine was assassinated in November 1902. The murder was blamed on the union, which did nothing for its increasingly violent image. Finally, in 1903, the truce collapsed and a general strike was called. Strike breakers, threats, violence—a familiar pattern emerged —and the owners reacted by asking Governor James Peabody to send in the National Guard. Before the troops were finally withdrawn in June 1904, unionists were driven from the district and civil rights were thrown out the window. The mine owners, in the guise of a Citizens Alliance, had broken the Western Federation.

Directly connected with the Western Federation's fight in Telluride was the struggle in Cripple Creek. Dissension in the state's major district had been building since the nineties, both sides doggedly determined to win. The owners still smarted bitterly from their earlier defeat. Trouble started in the mills in 1903, when the Western Federation tried to unionize the workers. A strike erupted and many of the Cripple Creek miners stopped work in sympathy, which finally forced management concessions. In August the mill workers went out again, this time for an eight-hour day. For nearly a year afterward, there would be little peace for Cripple Creek.

The mine owners organized an association whose avowed goal was elimination of unionized labor in the district. They appealed to Governor Peabody, who marched in the troops. Strike breakers were shipped in, professional gunmen deputized, and union members blacklisted. The union retaliated with violence of its own, and the district was virtually divided into two armed camps with the troops maintaining an uneasy, and not unneutral, peace. Terrorism finally evoked a declaration of martial law. The struggle had been elevated to warfare for only one goal, unconditional surrender. The owners' heavy-handedness, Peabody's support, and Adjutant General Sherman Bell's activities (he personally supervised operations), resulted in apparent peace by February 1904. The guard was withdrawn, but the cost was seizure and deportation of union members, "capture" of the pro-union Victor *Record*, intimidation of local courts, prohibition of street meetings, and confiscation of strikers' firearms. Western Federation membership was depleted and an uneasy calm settled.

Desperate and losing, the Federation turned to further violence—raw and brutal violence that shattered the calm. In the early morning of June 6, 1904, Harry Orchard, a professional terrorist, probably union-backed, dynamited the railroad depot at Independence, timing his deed to kill the most possible nonunion workers crowded on the platform; thirteen died and many others were injured. This ghastly event (blamed by the public on the union) made it possible for the owners to triumph completely. Back came the guard, martial law, wholesale arrests, deportation, wrecking of union halls, and harassment of union members by mobs. By midsummer the strike was broken and the union with it, although the Western Federation of Miners stubbornly refused to terminate the walkout officially.

With the failure in Cripple Creek and Telluride districts, and setbacks elsewhere in Colorado, the Western Federation lost all that it had gained. Its leadership, outspoken and prone to radicalism in speech and action, had pushed too hard and too fast, converting strikes for improvement of miners' wages and working conditions into struggles for power. Ownership responded and found that with a favorable state administration, which Peabody's unquestionably was, they could win not just the battle but the war as well. In the end everybody lost. Colorado acquired the image of a reactionary state, unionism suffered a serious setback, mining districts and camps were hurt, and the miner was left at the mercy of management, having gained from all his suffering little but a radical image.

Nor was there peace in the coal fields. The northern and southern fields increased production steadily as

115. "Sold my soul to the company store," said the popular song. The Colorado Supply Company store was a feature in the CF&I communities; hated by many for unfair practices, it was defended by the company. Company stores were typical of the coal towns, an anachronism in the hardrock camps. Courtesy the State Historical Society of Colorado, Denver.

116. This coal camp—crowded, dusty, and ugly—held together because of the jobs it offered. The regimented company housing at the upper left contrasts with the individuality of the rest of the scene. The residents of Hastings are gathered for what could be a parade, or a funeral. Courtesy the State Historical Society of Colorado, Denver.

the century opened. Colorado continued its 1890s role as the leading western coal state, a fact not lost upon the state coal mine inspector in this 1905–06 report. There was, he said, a large demand for "our product," which proved a strong incentive for investing capital in developing new mines and equipping those mines with "modern machinery." Though he spoke favorably of the progressive spirit of the day, there was unrest in the coal camps and down in the dark, dangerous mines.

Life and working conditions had not changed noticeably since the 1880s. Indeed, they had become even more dangerous, as more mines opened and more miners went into the earth. A few examples will suffice to show a trend. In 1905 there were fifty-nine fatal accidents, followed the next year by eighty-eight. A horrifying total of three hundred nineteen for 1910 stunned even the mine owners. The Trinidad area was the center of this holocaust—seventy-five died at Primero, seventy-nine at Delagua, and fifty-six at Starkville. These men who died in 1910 left behind one hundred sixty-three widows and three hundred three children. Excuses and blame-placing on carelessness and other noncompany factors could not cover up the obvious fact that coal mining, as being conducted, was dangerous and those safety laws that were on the books were not being enforced.

That more contemporary attention was not brought to focus on these conditions testifies clearly to the local and state political power of the coal companies and the fact that many of those killed were foreigners. The 1910 disasters claimed far more Hungarian, Croatian, Mexican, Polish, Italian, Russian, and Austrian lives than American. They died politically powerless and little known or appreciated by the majority of Coloradans.

They lived out their lives in company towns only slightly changed from those of an earlier day. The Colorado Fuel and Iron Company, part of the Rockefeller empire, had replaced railroads as the major coal operator, and its representatives were careful to see that their towns were tightly controlled. Their residents possessed few, if any, tangible rights. A miner's job and dwelling place depended upon his unprotesting compliance and long work days—the eight-hour day had no meaning here. He bought his food and supplies from the company store (CF&I conveniently established the Colorado Supply Company to handle its retail trade), worshiped at the company-sponsored church, if one were available, read books from the company's circulating library, and sent his children to the company-backed school. To make matters worse, these miners could not even get along with each other—Italians looked down on Greeks, who despised Poles—and so it went.

It was not that the coal companies were unprogressive; they were advancing especially in one area, the introduction of machinery. Without question the use of mining machines was an important factor in the production increase during these years. Unfortunately, they did not increase the safety factor correspondingly.

Into the wretched lives of the miners (one reporter called them little removed from "downright slavery") came a ray of hope—unionism. The Knights of Labor made some inroads in the 1880s and 1890s with varying effectiveness. The companies opposed them with all the methods at their disposal, which were considerable. With the Knights' demise, local unions came to the fore; lacking much strength, they proved generally ineffective in representing the miners' interests. A few of these affiliated with the Western Federation of Miners, but this hardrock union did not meet their needs. Finally, just at the turn of the century, the United Mine Workers made their appearance. The companies rallied against this threat. Miners were told they would be discharged if it was discovered that they had joined the union. Union organizers were beaten up, intimidated, arrested on trumped up charges, or driven out of the camps on no charges whatsoever. Huerfano County was perhaps the worst, it being virtually a fiefdom of the CF&I which did not hesitate to use any means to maintain its control.

Despite such obstacles the union made some headway, and the miners became more militant. Between January and April 1901, a strike to increase the amount of money the miners received per ton launched a tense three-year period in which strikes flared over working conditions, wages, and other issues. Results varied, but the eventual outcome—the triumph of the owners—was inevitable. There followed a lull that nevertheless included some local disputes; then a series of strikes in the coal fields, starting in Boulder County, culminated in the tragedy at Ludlow in 1914.

The United Mine Workers had faced the most trouble in the southern fields, where conditions were the worst. Headway proved painfully slow; the CF&I fought them every inch of the way. Finally, in September 1913, the Trinidad area miners called a strike, demanding a ten-percent wage increase, recognition of the union, and enforcement of Colorado mining laws, including the eight-hour day and health and safety regulations. Another point at issue was the freedom to choose one's own living quarters and trade somewhere other than at the company store. Both sides dug in, the Union sending some of its best organizers and leaders, including John Lawson and Mary Harris, the miners' "Mother Jones," an eighty-two-year-old Socialist whose age diminished neither

117. Coal mining was more dangerous than hardrock because of the nature of the mineral and its dust. This view of miners at Starkville illustrates the crowded conditions under which the men were often forced to work. A true hardrock miner would not choose to work in coal mines. Courtesy the Colorado State Archives, Denver.

118. When accidents did happen in the coal mines, they frequently involved gas, fire, and explosions. Tragically, safety laws were indifferently enforced despite miners' protests. This view shows coffins being taken to Starkville after an October 1901 accident. Courtesy the Denver Public Library, Western History Department.

119. Eventually mining conditions, low pay, feudal dominance of the companies, and depressing camp situations led to organization of the miners. Strikes followed. The classic confrontation occurred in 1913–14 when the miners moved into tent cities. This one, White City, was near Walsenburg. Courtesy the State Historical Society of Colorado, Denver.

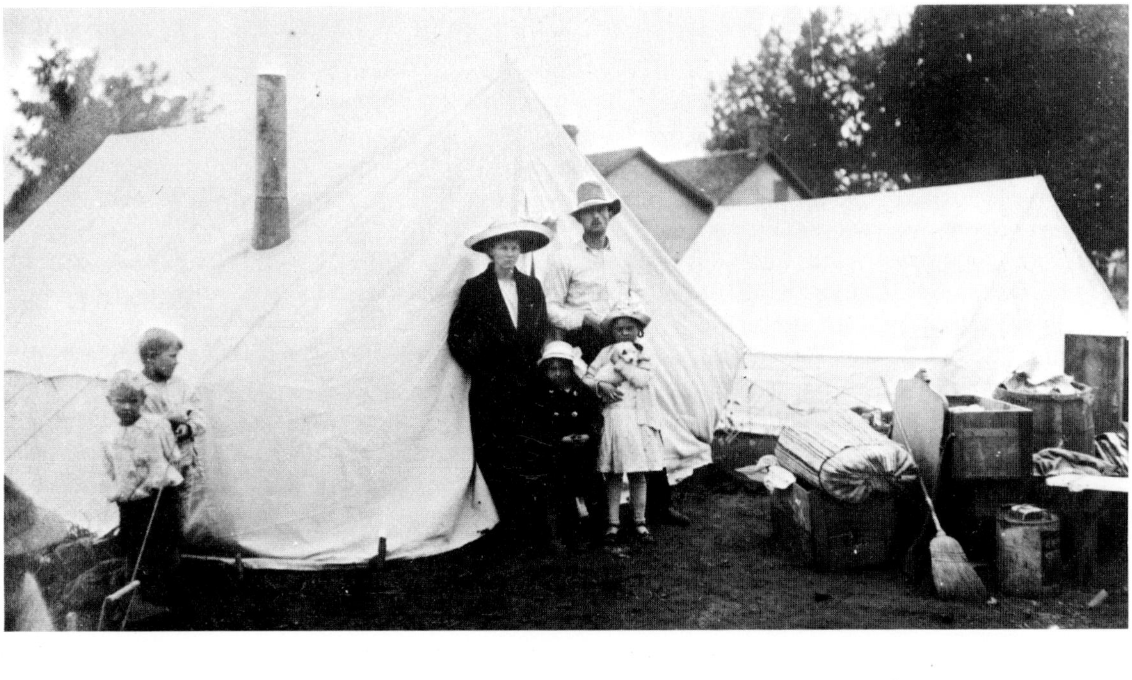

120. They don't look like devils without horns, but that is what strike breakers (shown here at Vulcan Mine, LaFayette) were to striking miners. Few men were so despised or the object of so much violence. The owners relied on them to keep the operations going. Courtesy the Denver Public Library, Western History Department.

121. When the owners turned to armed company guards to protect their property and the strike breakers, the miners were apt to arm and retaliate. This group shows the varied makeup of the coal miners. Despite their military appearance, they were pitifully short of real power. Courtesy the Denver Public Library, Western History Department.

her enthusiasm nor her dedication. The mine operators reacted with non-union labor and guards. Taunts, fist fights, and occasional sniping, plus many threats, marked the course of the strike.

Action begot reaction; tension and hatred festered. The striking miners, out of work and evicted from their homes, withdrew to tent colonies, sustained by union funds, near the mines. Hurried appeals to the governor brought in the National Guard, a now familiar oppressor to striking miners. Finally, on April 20, 1914, the inevitable happened. At Ludlow Station, north of Trinidad, miners and militiamen tangled, and the struggle spilled over into a tent colony. Five miners and one trooper died that day, as did two women and eleven children huddled in a hole trying to avoid the fighting. They died not from bullets, but from fire and smoke when the tents caught fire. Immediately the UMW labeled this the "Ludlow Massacre," and national attention focused on Colorado.

Governor Elias Ammons decided that the situation was beyond his control and asked that federal troops be sent in to replace the guard. An uneasy quiet descended; the strike dragged on, finally ending officially in December. Neither side won, nor did Colorado. The union was denied bargaining rights and withdrew from the state. A "company union" replaced the UMW, with a few short-range pluses that were overshadowed by the long-range weaknesses. The

122. Ultimately, the national guard was called in and faced the strikers across a narrow strip of land, reinforcing the owner's position. Such was the situation at Ludlow in April 1914. Courtesy the State Historical Society of Colorado, Denver.

123. The result could be needless fighting and tragic loss of life. Ludlow showed the bitterness and hatred that had built up; it also represented the nadir of management v. the worker in the Colorado coal fields. Coloradans could not forget what happened here and why. Courtesy the Denver Public Library, Western History Department.

owners, however, were forced to provide better working conditions, shorter hours, and a higher pay scale. They also received enough bad press to last a generation.

Coal production and value fell that tragic year of 1914 to the smallest total in a decade. The most bitter, violent, and bloody strike in Colorado's coal fields came to an end with an obvious dollar loss. But this was not the real loss of these strikes—the suffering, hatred, and deaths produced a wound that would not heal quickly. Coloradans would remember for years what had happened here and they would not be proud of it.

The smelter owners, who had triggered labor problems in several hardrock districts and were generally lumped with management in the union's eyes, had their ups and downs. Cyanidation, which had been tried first in the 1890s, pervaded Colorado smelting. Cyanide was found to be one of the few common, reasonably cheap, nonexplosive and stable compounds that had a high affinity for gold. Many of the pans, separators, and tables, which represented other methods, were replaced by large wooden vats. Stamps and crushers remained to do the primary work. Cyanide mills appeared in Cripple Creek, Florence, Telluride, Ouray, Leadville, and elsewhere. For a while chlorination held its own in Cripple Creek, but by 1911 it had given way and cyanidation was king of

124. The first years of the twentieth century were exciting ones for Colorado dredging. The Columbine No. 1 of the Tin Cup Gold Dredging Company, however, was sitting peacefully in its pond, its bucket line silent. Behind it, to the right, are electric poles, indicative of the new power source for mining. Courtesy the United States Forest Service, Washington.

the gold fields. There were still lead smelting plants at Durango, Salida (a new one blown in 1902), Pueblo, Denver, and Leadville. Various types of concentration and reduction mills were scattered about, many of them no longer active by the second decade of the twentieth century.

Smelting had taken a new tack with the appearance of the American Smelting and Refining Company, which came to dominate completely Colorado's industry. Consolidating its position, the company closed some of the less profitable operations, for which it was not forgiven for years. As late as 1934, in the *Annual Report* of the Colorado Bureau of Mines, it was blamed for the poor condition of smelting in the state, because it had closed six of the eleven smelters. Some of the blame might properly fall here; however, the labor troubles of 1903 had contributed to the closing of the Omaha & Grant Smelter in Denver, and declining production and profits weakened others. The most modern and best located survived. Transportation, as always, was the crucial factor, even if the low-grade ore could not stand the cost of shipment. Neither could the smelter stand the expense and worry of trying to maintain operations in a risky low-grade district. It was simply no longer feasible to have a smelter at everyone's doorstep. Smelting had become big business and would be run as such; no clearer indicator of the disappearance of the mining frontier was needed. Colorado smelters were now serving a sweeping area, including British Columbia, Idaho, South Dakota, Arizona, Nevada, and were even taking zinc from Kansas and Oklahoma, as well as home-mined ores. Colorado was only a part of the whole picture, however, a hard fact to swallow for local miners.

Behind American Smelting and Refining stood the Guggenheim family, which towered over Colorado and American smelting. Philadelphia-based Meyer Guggenheim and several of his sons had begun in Leadville mining, then branched out into what they felt was the more reliable end of the business. The outgrowth of their activities was the Philadelphia Smelting & Refining Company with plants at Pueblo and in New Jersey and Mexico. The Guggenheims did not immediately join the American Smelting and Refining in 1899, and a lively competition for ore resulted between the two. When they did merge, two years later, the Guggenheims received enough stock to

125. Dredging companies did not generally go to the expense of setting up an operation without some testing. This rig was testing gravel in the Breckenridge area before money was spent to build the dredge. Courtesy the State Historical Society of Colorado, Denver.

gain control and thus became the major influence in American smelting. The Tabors and Strattons of the previous generation had been replaced first by outside investors and then by industrial capitalists.

As significant to Colorado mining as the smelting changes was the more general acceptance of electric power. For example, in 1907 two companies furnished electricity to the dredges at Breckenridge. The next year electric power lines were completed into mining districts of Boulder, Clear Creek, Gilpin, Lake, and Summit counties. The San Juans, which had helped pioneer development, readily accepted the convenience and increased reliability of electricity in both mining and milling. Electricity made the miners' lives easier and brightened the nearby camps as well. It modified life patterns, allowing more night-time activities with greater comfort, and provided an added safety factor for homes and mines. Its warm glow lighted Christmas festivities and mine depths alike.

Besides accepting electricity as a power source, dredging companies were also busy on other fronts. In the first years of the twentieth century dredges clanked their way through many Colorado stream beds with varying degrees of success. The tentativeness of the nineties eventually yielded to high hopes and full-scale operations during one of the important eras for local dredging. Dredges at one time or another operated in what became Moffat County, and in Summit, Park, Costilla, and Routt counties. Breckenridge held center stage; during some seasons, as many as four dredges operated successfully along nearby streams. "Successfully" was the key word. High hopes and expectations came to be dashed on the rocks of reality, as represented by the boulders in the stream beds. A dredge operating on Placer and Sangre de Cristo creeks in Costilla County in 1910–11 finally overturned and was not righted again. Its owners were swamped by low returns and operating problems. Rocks of all sizes and shapes hindered Colorado dredging and shot costs beyond estimates nearly everywhere. Conditions had to be exactly right to sustain a profitable operation. Colorado never attained national significance as a dredging area, trailing Alaska, California, and several other states in total production and real interest in the enterprise.

Hydraulic mining also again came into vogue, its pressurized water streams shooting out of nozzles and eating away at river and hill bank. These were not

126. Boulder was more than a university town—it became the heart of an oil boom, as the Boulder field was "blown in." This was the McAfee oil well, with, most likely, its proud owners looking on. The oil industry by now was being run on a much more scientific basis. Courtesy the University of Colorado Museum, Boulder.

small operations. The Gold Pan Mining Company at Breckenridge owned nearly seventeen hundred acres and took its water out of the Blue River about four miles above the placer. Ditches and nearly two miles of steel pipe brought the water down against the gravel under a hundred and fifty pounds of pressure to the square inch. The company even built a flume to carry the Blue River's remaining unused water around the operations taking place in its recent bed. Two large derricks were used to remove boulders and heavy debris. The company set aside a dumping ground for the tailings in answer to a long-standing complaint against hydraulicking by those further downstream who found themselves inundated by muddy waste tailings.

Newcomers on the mineral block were uranium and vanadium. As far back as the 1870s, some pitchblende had been found on a dump near Central City and carnotite ore had been uncovered in southwestern Colorado. Not until the Curies finally unlocked the secret of radium, however, was a use for them found, and it was then that uranium became the center of some interest in western San Miguel, Montrose, and Mesa counties. Vanadium, for the moment, was of secondary interest.

Medical and scientific research were the prime markets, with the bulk of early production going to Europe. Paradox Valley in Montrose County became the chief producing area, with the Standard Chemical Company the largest single operation; in 1914 its mill shipped more than half of the total United States output. The federal government became involved when it joined with a group of doctors to secure radium. The National Radium Institute, acting under supervision of the Bureau of Mines, was active in the Paradox Valley and also maintained a Denver plant that produced radium and investigated reduction processes. A small amount of pitchblende continued to be taken near Central City, but nothing of overall significance. With a limited market, drastically shrunk by the outbreak of World War I in 1914, this industry was not in a healthy state as these years closed.

In better shape were two relatively unappreciated oldtimers. Zinc and copper, poor country cousins to the glamorous metals, finally gained some recognition of their own. Neither had seriously challenged the million-dollar production plateau until copper went over the top in 1896 and stayed there until 1914; zinc broke the barrier in 1901, reaching nine million in 1912. Copper was produced primarily as a by-product of precious metal mining in San Miguel, San Juan, Ouray, and Lake counties. Zinc was one of the few metals that showed a great increase, even if a fluctuating one, after 1900. Because Lake County was far out in front as a zinc producer, Leadville regained some of its lost glamour. By 1906 the county's zinc product was worth more than the total gold and silver output. Zinc-concentrating mills, then zinc smelters appeared. From being cursed as a troublesome ingredient in the smelting process, even to the point of penalizing high percentages of it in ore, zinc came to be touted as a new bonanza.

Oil, meanwhile, was having its troubles; production in Colorado had in 1892 reached a nineteenth-century peak of about 824,000 barrels, with Florence still the only producing field. A rapid and constant decline then set in, which more than halved production by 1900. Colorado's oil future looked bleak when the twentieth century arrived.

The success of a discovery well in 1902, in what became the Boulder field, gave rise to new excitement and initiated commercial production in this area. Stretching from Boulder to Longmont and beyond, other commercial wells soon joined the discovery, and a small refinery was built east of Boulder. The Boulder field reached its peak in 1908–09 with production in the 80–84,000-barrel range, less than Florence at its best. This older field at the same time stabilized and showed occasional years of increase.

By 1914 state production was again slipping; both the "old" Boulder and Florence fields were producing less petroleum. In northwestern Colorado oil had been found near De Beque and Rangely, a portent of what was to come. Unfortunately, at that moment, neither possessed adequate transportation facilities nor seemed rich enough to stimulate interest or generate needed capital. Colorado still had a way to go with oil before emerging as an important producer.

In the years from 1900 to 1914 a way of life vanished from Colorado, never to be replaced or even realistically relived; the mining frontier disappeared. It had not gone gently. The labor violence had ended the era violently, or, more to the point, had simply terminated what had begun a decade before. The frontier left behind a legend, which has grown since, becoming similar to one's first love, prettier and more alluring as the years pass.

These years witnessed something else as well. Gold mining peaked in 1900–02 (in the $27–28-million range), then gradually slipped into decline. Silver—in production, price per ounce, and total value—joined with gold, dropping under the $10-million level in 1902 for the first time since 1879. The precious metals were being replaced by lead, zinc, copper, and other ores valuable to America's new industrialism. Smaller operations and/or leasing were also coming to replace the large companies in all but the major districts. Change had come to Colorado mining.

TABLE 2
Top Five-Year Production of Colorado's Greatest Silver and Gold Camps

Cripple Creek-Victor	1898–1902	$81,882,268 (gold only)
Leadville	1879–1883	$50,080,851 (silver only)
Aspen	1888–1892	$28,879,215 (silver only)
Telluride	1906–1910	$16,234,038 (gold & silver)
Central City-Black Hawk	1868–1872	$11,770,957 (gold only)*
Ouray	1907–1911	$11,637,377 (gold only)

The price of ore is the average of each year; no attempt was made to base these figures on a standard price. For a camp to be tabulated, one or both metals had to average $500,000 per year, and a total of ten million dollars for five years. If the best five-year totals were ranked, rather than a single peak period, Leadville and Cripple Creek would monopolize the top five.

*Earlier production figures for Gilpin County are tentative at best, without placer production totals.

Source: Henderson, *Mining in Colorado*

9

Times Change, So Does Mining

The era of World War I and the years of the twenties were a time of transition for Colorado mining. Old minerals and veteran ways gave ground, even though a couple of old-timers finally came into their own and a new era was heralded with the coming to the forefront of some minerals answering modern markets' demands.

A general overview shows these years divided into two distinct periods: the war-stimulated boom, followed by a post-war depression; and the ups and downs of the 1920s. Military demands stimulated base metal production for the first time in Colorado mining history. A new experience then, it was one that would be repeated several times. Long before the United States went "over there," the needs of the Allied powers generated a spectacular market for some metals. Zinc was an example. Lake County's zinc total had multiplied steadily in the ten years preceding 1914, a fact obscured by the obsessive quest for the more glamorous precious metals. Large bodies of zinc carbonate ore had been found in nearly all parts of the district, and in the top pre-war year, 1912, over sixty percent of the county's ore value came from this metal. Production then slumped until 1915, when it zoomed upward to $10 million in 1916. Leadville continued as the major zinc producer in the state.

The impact of zinc on Lake's northern neighbor, Eagle County, proved even more spectacular. Isolated and somewhat forbidding, the county and its chief mining camp, Red Cliff, had languished except for a few brief spurts of activity way back in the 1880s. Then came the new reduction methods, better prices, and the war. Suddenly Eagle County was producing between one and four million dollars' worth of ore annually, with zinc accounting for almost ninety percent of that total. Long scorned by mine owners, and the bane of smeltermen because of the cost and trouble it caused in the separation process, zinc had now come of age.

Tungsten went through a similar boom, although in its case activity was centered basically in Boulder County. Like zinc, it had hampered gold and silver seekers and was frequently cursed as "that damned black iron." By 1900, it was generally known that this "black iron" was tungsten, and that it had a slowly growing commercial value. In the first decade of the new century, Boulder County emerged as the leading source, nearly eighty percent of the entire American output coming from its mines. Nederland, formerly only the mill town for Caribou's silver mines, now found its own fame as a tungsten "metropolis."

Recognized primarily for the extreme hardness and wear resistance of its alloys, tungsten was particularly valuable in high-speed tool steel, such as the cutting edges of dies and drill bits. These were vital compo-

127. Red Cliff was enjoying a zinc boom when this December 1917 photograph was taken. Eagle County finally had come to the forefront of Colorado mining. The large American flag shows the patriotic fervor of World War I. Courtesy the United States Forest Service, Washington.

128. Despite the 1920s slump, the United States Vanadium Mine near Rifle shows that the mining of this metal had come a long way during the previous decade and a half. Cars and trucks were standard equipment for the owner and miner. Courtesy the Colorado Mining Association, Denver.

nents in war-related industries and tungsten mining flourished, the price per pound skyrocketing from 45¢ in 1914 to $4.16 in 1916. Feverish activity disrupted Nederland and the northern part of the county in the scramble to find claims. It was an old-time rush, with cars and trucks replacing burros and wagons, and a movie theater on Nederland's main street. Claims were staked on every auspicious rock, holes became mines, mills mushroomed (Boulder had several) and Nederland's population soared from 300 to over 3,000. The rush passed quickly; by 1917 prices had dropped, and the miners and firms without much capital found themselves in trouble. The tungsten deposits were characterized by erratic distribution, with high-grade ore abruptly faulting and leaving only barren quartz. Many a miner found his seemingly valuable claim to be worthless after a thorough examination.

In 1918–19, the end came abruptly for tungsten when dull times in the steel industry caused the price to collapse and cheaper Chinese tungsten captured the available market. The downfall was hastened by exhaustion of the easily mined surface ores (quite often the richest) and the mounting expenses of underground mining. By 1921, Colorado and American tungsten mining were at their lowest point; a government report stated no ore was produced that year. Boulder County accounted for most of what was mined during the rest of the decade.

The rise and fall of tungsten paralleled much of the rest of Colorado mining. Better prices and increased demand temporarily reversed the downward cycle of gold, silver, lead, and copper early in the war, but in 1917 it began again. Gold, especially, suffered a disastrous slump. Reports by the state's mining inspectors give the reasons: a labor shortage, as men went off to war or war industries, the higher cost of explosives and mining supplies, a shortage of chemicals, and increased freight and smelting charges drove many out of mining, or, at the least, checked production. Even the influenza epidemic that swept the country in 1918 had its effect. Nederland and vicinity, for instance, recorded about forty deaths, and few mines or mills in Colorado worked at full capacity.

This war-stimulated boom created more long-range problems than had been anticipated. Forced extraction and undue speed had supplanted systematic development and steady shipments. Far too often little, if any, attention had been paid to maintenance and development of ore reserves, resulting in depleted mines after the armistice. In the mining depression of the late 1910s and early 1920s, Colorado could not compete and found itself dropping behind. It would be 1922 before an upswing was noticeable.

Unexpected, though not unusual, side effects caused further troubles. In this case, it was the closing of the smelters, which were hard hit by the production decline and low metal prices. Denver's Globe smelter was one of the first to go (1919); it was followed by the Salida smelter and then Pueblo's, which had been damaged by the famous 1921 flood. This left only the Durango and Leadville smelters operating. In 1927 a commentator asserted that these closings produced a far more disastrous impact on the mining industry than all the ill effects that arose out of the postwar depression. In spite of the fact that the railroads fixed reasonable freight rates to remaining plants, the small or marginal operator in Boulder, Clear Creek, or Gilpin could not withstand the increased transportation costs. Neither could his counterparts very far from Leadville or Durango. The smelter shortage plagued Colorado for years to come.

Not all mining was benefited, even briefly, by war-activated demands. Europe, the most important buyer, stopped its carnotite purchases in 1915, nearly killing the market. The fledgling uranium industry was sent reeling and did not recover for the war's duration. Stopping to analyze the situation, John Mullen, field manager for the Standard Chemical Company, put his finger on a problem that had vexed this particular industry since its inception. Carnotite mining was not very profitable for the poor man, experience having shown that, in order to make a profit, "one must go on through with it and sell the finished product, such as radium, uranium and vanadium." The open market paid very little for the ore itself. This was not the game for the small-time miner.

Other factors also weighed against him. The ore pockets were shallow; one that contained a hundred tons was considered good. The sites were isolated from the railroads, over narrow, steep roads that were infamous for their sharp curves. It was nearly as economical to use horses and mules as trucks, because of the road conditions, and as a result freight rates remained high. The aridity of western Colorado's plateaus and canyon lands bedeviled these miners; it was not unusual to haul water between one and three miles to some camps, thus putting a damper on the Saturday night bath. Faced with primitive, difficult conditions and no market, the industry stagnated. The lure of easy profits still tempted a public that was gullible enough to buy attractively presented stock. Until investigation proved the pitchblende deposits to be nearly non-existent, the Colorado Pitchblende Company, operating near Jamestown in Boulder County, busily offered its stock in full-page newspaper advertisements (in Denver and Salt Lake City) in 1918–19.

A brief recovery in 1919 soon evaporated and only glutted the unstable market. Nothing went right for the Colorado miners. The opening of the extremely

129. Molybdenum came into its own in the 1920s, but the Climax Company still had to confront problems of elevation and weather at its company town, Climax. A windy, cold January day in 1929 illustrated what winter meant; the swirling snow blotted out the mountains in the background. Winter took its toll regardless of the year. Courtesy the State Historical Society of Colorado, Denver.

rich and cheaply produced Katanga uranium deposits in the Belgian Congo stripped them of their world and national markets. The mines closed one after another, until by 1927 no ore was mined for uranium or radium. Near Rifle, the Union Carbide Corporation mined and milled vanadium, but overall the industry languished. Miners in several former uranium and radium districts of southwestern Colorado turned to vanadium, a product that had a better future.

Before it appears that these years produced little but fleeting hopes and lasting problems, attention should be turned to several minerals which prospered. One of these was new, molybdenum, which, like tungsten, was used in the manufacture of high-temperature steel alloys and castings. Molybdenum, when combined with iron and steel, imparted a hardness, toughness, and resistance to wear and corrosion. Molybdenum did not inspire the same rush as tungsten or gold. Not until 1900 was it identified, and then ten years passed before some use was found for it in industry. Colorado was perched on the largest deposit of this rare metal in the United States, if not the world, at what became Climax. Uses were found for molybdenum, and war brought the demand that opened the Colorado mines. Mining had been carried on in and around Bartlett Mountain (Climax) for years in search of gold and silver, ignoring the molybdenite that interfered with the work at hand. Finally, in 1915–16, interest was such that action was taken.

Through a series of consolidations and purchases, the Climax Molybdenum Company bought out many of the smaller interests and began full-scale mining operations. It was not done quite that smoothly, however; charges and countercharges of false entry, claim jumping, and the like occurred for years and were bitterly fought out in the courts. Once the value of the deposit became known, even Lake and Summit counties argued over the spoils. A vaguely drawn boundary now became of prime importance. By 1918 two large companies remained, the aforementioned Climax Molybdenum and the Molybdenum Products Corporation. The former eventually absorbed the latter and monopolized the district.

Among the immediate obstacles faced were having to work at elevations as high as 12,000 feet and climatic conditions described rationally as harsh. To combat these hardships, the Climax company quickly built boardinghouses at the mine and mill, a hospital, and tried to provide some recreational facilities at its isolated location. Nevertheless, it suffered a high labor turnover. In the beginning this did not seem so serious: excitement grew over construction, mine development, and the first shipment of concentrates which went out in February 1918, over a spur line of the Colorado and Southern railroad. Gone were the days of a "blast and a prayer"; this was big business that demanded sound financial backing, trained management, and national and world marketing facilities.

130. One of the most unusual of the Colorado mining operations was at Marble, where blocks were quarried for the Tomb of the Unknown Soldier and the Lincoln Memorial. This was the workshop of the Colorado-Yule Marble Company, where the men were cleaning and polishing the marble. The finishing mill was once the largest in the world. The town of Marble went through several fluctuating economic cycles before the quarries closed in 1941. Courtesy the Denver Public Library, Western History Department.

The opening of Climax stirred interest in molybdenum. Soon other Colorado deposits (61 according to a 1919 report) were found. Chaffee, Gunnison, and Summit counties led the parade, but most of their ores proved not to be marketable commercially. Only one, the Camp Urad or Urad Mine, in Clear Creek County above Empire, sustained long-term interest and production. At the moment, the Primos Chemical Company operated it on a smaller scale than Climax.

Despite the rosy promises, molybdenum was caught in the post-war slump, and Climax ceased operations in April 1919. Camp Urad closed in July. A plunging price (down to 72¢ a pound from an earlier high of $5) and a large surplus stopped mining cold. A 1919 writer felt the outlook was none too bright and "what the future has in store no one can say." The Climax Molybdenum Company had something to say; it kept its plant in a standby condition and set about strongly promoting an advertising campaign to extol the value of molybdenum steels. It kept itself afloat by selling the stock-piled concentrates to the small available market. But the molybdenum situation deteriorated, nevertheless, and, like tungsten, no production was noted for 1921. Unlike tungsten, however, molybdenum was, at this low point, turning the corner; it was the darkness before the dawn of bonanza.

The steel industry was making a slow comeback after its post-war traumas and, with it, the use of molybdenum. America's growing love affair with the car propelled the industry to the status of an industrial giant, and molybdenum, alloyed with steel, was used for axles and springs. The Climax Molybdenum Company's advertising campaign paid off when it succeeded in getting orders from automobile manufacturers, particularly from Ford. The mine and mill reopened in 1924 and, within a year, several steel companies were purchasing carload lots.

Upon reopening, the company faced the same personnel problem as before—a labor turnover so great that by the 1930s it had one of the country's worst records for an industrial operation of its size. A major construction program to counteract this attrition was launched at the town of Climax (a company town—twentieth-century version), which produced well-built homes with electricity, running water, and many modern conveniences. Still there were never enough homes. A recreation hall and even a ski run provided diversions. Miners continued to drift in and out, however. Most newcomers suffered from headaches, drowsiness, and other complaints associated with high altitudes. Even after acclimation they found their efficiency impaired. Another factor contributing to the transience was the social isolation, which particularly affected the single miner. He could go to Leadville, the nearest town, where "vice of all kinds" was available for a price. The company found that such visits adversely affected its workers, eventually cutting into the miners' working time and eroding efficiency.

131. Coal production peaked during and after World War I. Many areas of the state had coal mines, including the town of Cameo, near Grand Junction. The tipple and mine buildings belonged to the Grand Junction Mining and Fuel Company. Courtesy the United States Bureau of Reclamation, Washington.

Leadville, well past its 1880s peak when it had proclaimed itself the nation's leading red-light district, still maintained its allurements.

Molybdenum emerged in the 1920s as one of Colorado's important mineral products. Never as colorful, and corporation-controlled from the beginning, it did not catch the public's eye, but the shrewd investor could harvest a financial windfall, provided stocks were purchased at the right time.

Coal proved to be another bright spot; after years of gradual growth, it finally emerged into the limelight. The industry rebounded quickly after the strikes and violence of 1913–14, unimpeded by any labor disturbances in 1915. The next three years were even better; 1917 set an all-time record high of 12,483,000 tons mined. This production coincided with the war years' market, when there was an unprecedented demand for coal in the iron and steel industries and increased call for coke in the base metal smelters. Over 14,000 men worked in the coal fields during these prosperous years. The coal they mined was used by railroads, local industries, and for domestic purposes in Colorado, Kansas, Nebraska, and Texas, as well as in the war-related industries.

Colorado was the largest coal-producing state west of the Mississippi River. Her two big fields, northern (Adams, El Paso, Boulder, and Weld counties and points in between) and southern (Huerfano and Las Animas counties), continued to be the major centers. The Canon City, Yampa (Moffat and Routt counties), and southwestern coal fields (La Plata and Montezuma counties) along with various individual mines, were the other producers. There was no question of which area ranked first—it was the southern coal field, with Las Animas and Huerfano the first and second statewide leaders.

Such a pace of production could not be maintained once the war pressures were relieved, but there was no major slump here as there was elsewhere. Coal production leveled at the ten-million-ton range in the 1920s. By 1929 coal producers, long scorned as second-class mining citizens, could proudly point to the fact that the annual output of coal exceeded in volume and value any other mine product. That year it also ranked first in aggregate value. Gold and silver had been dethroned.

Underneath the surface prosperity, all was not well in the coal fields. Already oil and natural gas, for industrial and domestic use, were making inroads into coal consumption, and 1927–28 proved to be another period of labor unrest. Again wages and working conditions were at the root of the conflict. It still was less safe to work in a coal mine than in the hardrock mines, as the accident and death rates continued to show. The worst clash between striking miners and mine guards occurred at the Columbine Mine near

132. This crew stands in front of its coke oven at Segundo about 1920. Coke continued to be an important product of coal, and neither the work nor the ovens changed much over the years. It was still a hot and dirty job. Courtesy Colorado Fuel & Iron Steel Corporation Archives, Pueblo.

Lafayette in November 1927, and resulted in six deaths. Martial law was promptly declared, and the strike was broken.

Coal and molybdenum were not about to arouse much excitement during the "roaring" twenties, an appellation inappropriate for most of the mining districts and towns. If the twenties roared for anybody, it was the urbanite; he was caught by the glamour of easy living and stories of quick wealth, and Colorado provided him with a bauble which attracted and kept his attention—oil shale. Here was something "practical," something that could propel his new plaything, the automobile, and do other wondrous things. If the advertisements could be trusted, it had been found practical even for "the relief and checking of the spread of social diseases." Oil itself was rather old hat, having been found in Colorado decades ago, but oil shale was new and exciting and seemed to hold the key to easy wealth for those lucky and wise enough to invest.

Oil shale had been fascinating people since the 1880s. According to local legend, one early pioneer of the Rifle area put up a fireplace made of a local dark, easily worked rock. He had an unexpected housewarming when he lit the first fire in it. In 1906, oil shale was tested as a road-building material. Finally, in 1913–15, an examination was made to determine the richness of Colorado shale, out of which came an optimistic 1916 government bulletin. Demand for the pamphlet was such that the edition was soon exhausted —and the rush was on. Caught up in the fervor, one writer bravely forecast: "The recovery of petroleum by the distillation of shale promises to develop into an important industry in western Colorado in the course of the next four or five years." This was a forerunner of many similar statements.

Interest centered in Garfield, Rio Blanco, Mesa, and Moffat counties in the northwestern part of the state. Overnight, De Beque (Mesa County) became the new Mecca and also a center of retorts for treating the shale. Companies were organized, claims staked, and various retorting experiments conducted. Oil shale had been mined in Scotland for years, and it was assumed, in the urgency of the moment, that the same treatment methods would work. Up went "Scottish" retorts, to prove only that oil shales were different the world over; foreign processes proved futile for the bulk of Colorado shale. The all-American nativists were secretly pleased, no doubt, that American oil shale would not fall prey to foreign technology. The shale oilers felt differently, however, because this put a definite crimp in their plans.

The process that finally evolved during this phase of oil shale madness was based on heating the shale to generate and evaporate the oil, which was then recovered by condensation outside the retort. According to an intoxicating rumor which raced through Colorado, the value of oil shale would exceed either

133. To extract the oil from shale, a distillation process, which featured a retort, was devised. When combined with pipes, tanks, and other gadgets of the inventor, the operation was ready to go. This group undoubtedly shows both the workers and owners all set to begin, probably with someone else's money. Courtesy the United States Geological Survey, Denver and Washington.

the total yield to date, or that which might possibly ever be extracted, from all the state's metal and coal mines. A 1918 estimate placed the reserves at twenty trillion barrels. It staggered the imagination.

Oil shale stocks and claims now came to be the most sought-after in the mineral world; by 1920, there were well over a hundred oil shale companies of various competence mining the stockholders' pocketbooks, if not the shale. De Beque, Grand Valley, Rifle, and Grand Junction all felt the excitement, and even a *Shale and Oil News* was published. It was claimed that oil shale gasoline produced thirty percent better mileage than traditional petroleum, that it could be used to manufacture soaps, fertilizers, mineral oil, dyes, sheep dip, fine harness dressing, and automobile paint. Almost anything, from oil to medicine, could presumably come from this new miracle product.

To insure an ample supply of oil, presidents Woodrow Wilson and Calvin Coolidge set aside reserves in Garfield County in 1916 and 1924. Some operators thought this smacked of socialism, but they need not have fretted. There was plenty for all, and no one had as yet found a process able to handle shale at a sufficiently low cost to justify commercial development. It could be done—some investors had purchased stock after watching a small mobile retort smoke up their neighborhood. The cost factor continued to be crippling. By the mid-twenties the excitement was over, the boom dead, and northwestern Colorado returned to a more normal existence. Many stockholders were left holding ornate stock certificates rather than the dividends they envisioned. As for that miracle product, it had been derailed by such practical considerations as cost, the sudden growth of California, Texas, Oklahoma, and even Colorado oil fields, and by a depressed crude oil price, well below that of shale oil.

The government did not give up. At the suggestion of a special presidential committee, money was appropriated and construction started in 1925 on an oil shale pilot plant near Rulison. The state had been doing its bit by opening and maintaining an oil shale laboratory at the University of Colorado. The point of these efforts was to test shale in a small way to determine costs of mining and recovering oil. Forecasts were made that a new industry, comparable in scope to coal mining, would be developed in the near future, but the Rulison plant was not able to resolve the cost impasse and it ceased to operate during the depression.

One of the factors which doomed oil shale development was the growth of Colorado oil fields. Though it was one of the oldest oil producing states, Colorado had not for years registered any significant amount (.07% of the country's total in 1916, for example). The Florence and Boulder fields were still the most productive, with scattered wells elsewhere. Wildcat drilling was not discouraged and was spotted around such distant towns as Fort Morgan, Flagler, Pueblo, Aurora, La Junta, and Steamboat Springs in 1917. Until 1923 production faltered dismally—no relief was apparent. Then, within two years, oil was discovered at what became known as the Moffat and Fort Collins fields. A new era in Colorado oil opened.

Colorado's production climbed from 86,000 barrels in 1923, to 1,226,000 in 1925, and on up to a high of 2,750,000 in 1928. This did not indicate only the effect of new wells; the older fields, Florence leading, managed to increase their production, while the search for new areas continued (twenty-three counties had test wells in 1924). As the derricks went up, other

134. Gushers were familiar scenes in Colorado in the 1920s, as well after well struck oil. Though unidentified, this Colorado field was no wildcat operation; the number of derricks indicates previously successful drilling. Courtesy the Continental Oil Company, Denver.

necessary accouterments appeared, such as pipelines and storage tanks. Local counties received an encouraging boost because of more jobs, taxes, increased trade, and a broadened economic base. But there were new problems to cope with, too: more schools, roads, and services had to be provided, and oil people were not always compatible with their agricultural neighbors.

Paralleling these developments was the emergence of natural gas wells. In 1912 a report had noted that, aside from the small volume of gas obtained from oil wells used locally in the Boulder and Florence fields, Colorado's natural gas came from individual wells, each furnishing the domestic needs of only a few families. Such small-time production ended in November 1923, when the Wellington dome (Fort Collins field) was brought in as both an oil and gas field. This was followed by others in Larimer, Moffat, and Jackson counties and the opening of markets via pipelines to Utah, Wyoming, and numerous Colorado cities.

At the moment, these developments offered only a taste of what was to come. Colorado's cup, though not overflowing, was certainly full. Mining of precious metals had declined, and uranium and base metal production was following a fluctuating course, but oil, molybdenum, and coal had surged to the front and oil shale had provided excitement reminiscent of another time.

Colorado mining was suffering from problems as the decade drew to a close. Some of the old districts were still active, having survived the vicissitudes of the years since 1914. Cripple Creek remained the leading gold district and was reported to have prospects that were "very bright." Gilpin County was "very active." Gunnison County still held bright promises after all these years: "a great part of this mineralized expansive area has been imperfectly prospected, and but a fraction of its mineral resources disclosed." It sounded like the 1880s again. Silver faced a harder time; concerning Aspen, one commentator mentioned that it was not "in the hey day of its youth," but it continued to produce fair tonnage. When the Pittman Act had been in force, from 1920 to 1923, silver activity abounded. The government-guaranteed price of $1 per ounce for domestic silver producers (to replace bullion sold for war purposes) had stimulated production. When it ended, the price dropped by a third within a year. Silver slumped again.

During the years covered by this chapter, the number of mines (gold, silver, copper, lead, and zinc) had decreased sharply—from 748 in 1917 to 359 in 1928. Both large and small ones shut down. San Miguel County was perhaps the hardest hit: all of its major mines, including the Tomboy and Smuggler Union, closed. Ore values were not high enough to sustain successful operation of these older properties, some of which had survived thirty years of steady production.

135. Prospectors and miners still burrowed away in the hills looking for gold and silver. The "We Got 'Em" lode did not "have 'em" despite the dog who stands forever threatening the challenger. This new operation showed that neither the interest nor the drive had evaporated in such old mining areas as Gilpin County. Courtesy the Denver Public Library, Western History Department.

Even with all the strides made in transportation and mining technology, Mother Nature could still throw her weight around, as Colorado miners found out again in 1929. An unusually wet spring that year caused flooding and mud slides, virtually stopping mining in the San Juans for months and curtailing ore shipments in almost every district in the state. Man also contributed to disrupting mining. The proposed junking of the South Park branch of the Colorado and Southern Railroad so unsettled mining along its route that several mines suspended operation and a couple of "big deals" were lost before the sale could be completed. Then came the crash of 1929; its jolt was just beginning to be felt as the year ended. A couple of years later, the Colorado Commissioner of Mines, John Joyce, was moved to observe that the crash might not have happened had the great part of the country's wealth gone into mining rather than the "coffers of predatory wealth."

Optimism prevailed, even though the mining frontier had passed nearly a generation ago. Its heritage lingered. In 1929, the commissioner of mines could still say:

> There are many well-known heavily mineralized sections of the state that have never been thoroughly prospected because of their early date remoteness which, at the present time, with modern advancement made in the means of transportation, coupled with new metallurgical discoveries and inventions, offer promising and fruitful fields.

10

Depression, War, and Uranium Fever

The depression years conjure sharp images, even decades afterwards, for the generation of Americans who endured them. For Colorado mining, the picture only got gloomier, since the previous decade had not been one of general mining prosperity. The 1930s provided little relief. The first three years were the worst, as they were for the state in general. The price of silver sank to new lows, reaching 38¢ per ounce in 1930. Some analysts thought the bottom had been reached, but by 1932 silver was down to 28¢ per ounce, lead to 3¢, and copper to 6¢ per pound.

Reacting to the deteriorating situation, the Colorado Commissioner of Mines, John Joyce, pointed out in his 1933 report what he believed to be the crying "needs of the hour" to restore the mining industry to the commanding position it had occupied in former years. A familiar chord was struck when he called for well-organized, coordinated, and modernized milling, smelting, and final reduction systems at one or more convenient, central locations. These would be supplemented by mills for concentration in the outlying districts, which, when put into operation, would permit the profitable treatment of all ore grades. This would, in turn, encourage the reopening of mines idled for years because of low-grade ore. Joyce also called for the "courageous" development of newly discovered mines and prospects; given the price of most metals, such a risk would have required considerable courage.

Financing, as usual, caused the most trouble. Money was not available internally, and outside capital was just as scarce. Joyce and the others knew this, but the commissioner went one step further in rebuking the attitude of the local business, industrial, and financial communities. What was needed from them, he wrote, "is intelligent and loyal support in a wholesome manner, instead of the disheartening way that generally marked their attitude in Colorado during the past decade."

Statistics, the dry but invaluable comparative tool of the economist and historian, show clearly what happened to much of Colorado's mining industry during this decade. Coal production declined twenty-three percent, lead forty-eight percent, natural gas forty percent, and zinc a shattering eighty-six percent. Petroleum slumped, too, the wells drilled numbering only about half those sunk from 1925 to 1930. On the bright side, gold, silver, and copper output improved, but the most remarkable of all was molybdenum mining. The Climax Molybdenum Company expanded steadily during the thirties, keeping Lake County in the forefront of Colorado mining.

The Climax Mine, Colorado's largest operation, was becoming internationally famous; in 1937 it produced seventy-one percent of the world's and seventy-seven

136. The Climax mining operation was the state's largest when this photograph was taken in 1939. For over a decade Colorado had been the world's major molybdenum source, a position not challenged until after World War II. Climax remained a prime example of a progressive company town, offering services and conveniences not available in many mining communities. Courtesy the United States Forest Service, Washington.

percent of the domestic output. From five million pounds in 1933, production quadrupled to over twenty-two million in 1937. Hampered by a shortage of skilled miners and machinists, the company opened a training school, probably one of the first in the state. It also sponsored a continuing construction program that provided jobs unrelated to mining and created such results as an enlarged mill and more employee homes.

For the gold and silver miner, all was not gloom, even though molybdenum was outshining the precious metals. The government rode to a partial rescue with President Franklin Roosevelt's sweeping New Deal programs. The American people were ready for a change in 1932. The Democratic Party and its candidate, Roosevelt, seemed to offer that, although just what they planned to do was vague. After Roosevelt's inauguration in March 1933, the country and Colorado soon found out, as pronouncements and acts dealing with a multitude of problems flowed out of Washington and, in the process, created a jumble of governmental agencies.

Few of these affected mining directly, but those that did had an impact on Colorado. When Roosevelt came into office, the price of gold had been pegged at $20.67 per ounce for years. As part of his program, the president asked in January 1934, for the Gold Reserve Act, which would authorize an increase in the gold price. It passed within a matter of weeks, and Roosevelt raised it to $35. The western silver interests, meanwhile, had not been dormant. With their old agrarian allies, they pressured the government for help in 1933–34. Out of their demands emerged the Silver Purchase Act of 1934, authorizing the treasury to purchase silver and to put into circulation silver certificates, redeemable in silver dollars upon demand. Some of the silver spokesmen of the 1890s would, no doubt, have been pleased with the success of their lineal descendants.

The old Bryanites would also have been pleased when Roosevelt, for all practical purposes, took the country off the gold standard through a series of executive orders and legislative enactments in March and April 1933. Gold coinage was abandoned and gold exports prohibited without treasury consent; Coloradans found that they, like the rest of their neighbors, could no longer own gold, except in a few specified cases. Of more immediate importance to the miner

137. These ladies are learning the art of placering in a mining class sponsored by the Public Works Administration in Denver. A fifty-niner would likely have turned over in his grave, but times were bad. They did not have to go far to try their skills, because gold was being panned in Denver itself, Arapahoe and Adams counties, and in the mountains. Courtesy the Denver Public Library, Western History Department.

138. More productive than the ladies were these operators in California Gulch in 1931. Placer mining prospered in the 1930s, as it had not for decades, reflecting increased gold prices and hard times. The third best year for ounces of gold recovered since 1868 was 1941. Courtesy the Colorado Mining Association, Denver.

139. The old order was changing. Fire, weather, and dismantling were obliterating abandoned mining equipment and buildings from the Colorado scene. This stamp mill was being dismantled, a process that was stepped up noticeably when war broke out in 1941 and created a need for scrap iron and other metals. Courtesy the Western Colorado Power Collection, Center of Southwest Studies, Durango.

and his mine was the government announcement in 1933 that the yearly assessment work, required to insure continued ownership of claims, was temporarily suspended. This remained in effect until 1939.

As a result of these governmental actions, interest in gold and silver mining mounted. In 1934 the number of active lode mines doubled, a clear boost to the industry. Gold was more attractive for the same reasons it was in 1859—it could be found in the natural state and recovered without expensive equipment or great skills. For the first time in years, placer mining became extremely popular. Men who had never handled a pan or sluice suddenly took it upon themselves to acquire the necessary skills and venture out to the streams. Interest centered in the Denver area and surrounding counties, and on Cherry Creek and Gregory Gulch, where it had all started so many years before. In such bygone areas as Russell Gulch and north Clear Creek, and in newer ones like the Happy Canyon placer in Douglas County, men shoveled and washed. They persevered, even though the time consumed exceeded the meager cash returns, because few employment opportunities were available. The increased gold price also encouraged them to work, and many a man panned away the summer months making at least enough to keep himself and his family, if any, fed. The Denver region was a natural lure because of convenient markets for small quantities of gold (only two ounces or more had to be sold to the mint, less could be purchased by dental supply firms, jewelers, and the like), the large number of unemployed in Colorado's one urban center, and an almost forgotten reputation as a placer district. Placer operations, in fact, were carried on within the city limits. On a larger scale, dredges abraded the river beds in Park and Summit counties. One dredge in the Breckenridge district handled 808,000 cubic yards during the 1937 season, its eighty-eight buckets probably moving more earth than its human counterparts did during the whole decade.

The rise in the gold price also helped to reactivate lode mining, as did the reduced cost of materials and supplies and a labor surplus which pushed wages down. These conditions were symptoms of the depressed situation. Cripple Creek continued to be the largest gold producer. Water hampered operations considerably when the works dropped below the Roosevelt drainage tunnel. In July 1939, the Golden Cycle Corporation, which operated mines in the Cripple Creek district and a mill at Colorado Springs, planned and started a major engineering project, the Carlton Tunnel, to drain the district to a lower depth. When it finally was opened in 1941, at a cost of over a million dollars, Cripple Creek was caught up in war-related mining problems, which delayed the anticipated benefits.

A few gold sparks rekindled memories of earlier excitements. The Red Arrow discovery in 1933, on the western slope of the La Plata Mountains in Montezuma County, was one of these. A "rush" of prospectors scurried about staking claims and for the rest of the decade continued to mine. An occasional shipment of high-grade gold and silver ore justified their efforts.

Silver never bounced back as did gold; in fact, the Eagle Mine (and Eagle County), a combined copper-iron-gold-silver operation, was the leading silver producer for most of these years. Trailing badly were three San Juan counties—Mineral, San Miguel, and San Juan—and the rest of Colorado reported only spotty silver production.

The smelter shortage, as Commissioner Joyce had indicated, proved to be a critical one; the industry suffered another setback from a court case instigated by Clear Creek Valley farmers. In 1935 the Colorado Supreme Court handed down a decision prohibiting the dumping of mill tailings into streams. "A primary duty rests upon one introducing such extraneous matter into this stream, to prevent damage arising from such introduction either from his acts alone, or in conjunction with those of others." The question of pollution and injury to streams was far from solved, but to the Colorado smelting industry another barrier had been raised (regardless of how necessary to the environment) against regaining the position it had lost after the turn of the century.

Mining accidents plagued the industry. Colorado in 1931–32 held the dubious claim of having the highest average number of accidents of any state. The state's Bureau of Mines continually pushed for stricter observance of safety laws and departmental regulations. The campaign paid off, and by 1939 the Bureau was congratulating operators and employees on their efforts to lessen accidents. Not yet completely satisfied, the commissioner noted that progress still needed to be made.

As the thirties ended and war broke out in Europe, Colorado mining still faced an unsettled future. Taxes were rising, foreign investors were being cut off, base metals prices remained low, and even such things as the high cost of workmen's compensation insurance worried some operators. The old-time mining speculator wishing to peddle stock also found the going rough, as the Securities and Exchange Commission clamped requirements on stock issues; no longer would "gold in the sky" sales pitches be so freely allowed. Long overdue as a buyer protection, such regulation nevertheless hurt some operations. These shady promotions had occasionally furnished money for a mine that turned out almost as promising as the promoter said it would.

The Colorado mining recession, which dated back to the late teens, eased as the United States geared for war in the early forties. The war years were a watershed and in the late 1940s mining again pushed Colorado into the national spotlight. The war years themselves were more than just transitional; they left a mark that has not been erased completely in the decades since.

Since the days of the early New Deal, government regulation had been a fact of mining life. But it was nothing compared to what transpired from 1942 to 1945, when miners could hardly swing a pick without filling out a form or complying with some regulation. The most infamous act passed during the entire war (in the opinion of some operators and mining communities) was the Gold Limitation Order, L-280, issued in October 1942. It closed nonessential mineral mines by

140. Base metals, such as zinc, were especially vital and the New Jersey Zinc Company near Red Cliff was one of those that prospered throughout the 1940s. Because of wartime restrictions and by-products of mining, Eagle County was Colorado's number one producer of gold, silver, copper, lead, and zinc in 1945. This photograph shows Belden, the shipping point of New Jersey Zinc's Gilman Mine. Courtesy the New Jersey Zinc Company, Gilman.

not allowing operators access to replacements or materials and forbidding work other than maintenance. Dredging operations ceased immediately, and gold mining gradually closed down; Cripple Creek was given a reprieve until June 1943. The purpose of L-280 was to free miners and equipment for the mines that were producing more critical metals, such as copper.

This order did not, by itself, suddenly put a stop to Colorado gold mining. Early in 1941 men began leaving the mines to tap the higher wages found in defense plants. Late in the year, operators were facing equipment shortages, which, along with increased taxes and the fixed gold price, pushed their backs to the wall. In March 1942, mines which produced gold and silver in excess of thirty percent of their total output lost their priority ratings for obtaining essential supplies and equipment. Order L-280 just completed a process that was already well underway, and it displaced about 2,700 workers throughout the country by the year's end.

The government could dictate what miners could and could not buy and whether they could operate. But the regulations did not end there—all phases of mining found themselves under government control. Numerous wartime boards were established, including the War Manpower Commission. Among the policies instituted by this agency were ones giving occupational deferment to nonferrous miners (former gold miners were assigned only to nonferrous mining jobs—except in undue hardship cases), prohibiting miners from changing jobs without government approval (one could move, but he lost his classification), and paying transportation of western workers, their families, and household goods to nonferrous mining districts. To accommodate and encourage movement, the commission, in November 1942, entered into the home construction business and offered increased pay to men who migrated to states with labor shortages.

Even with all these efforts, the shortage of miners continued, and the war department was finally forced to furlough 4,300 erstwhile soldier-miners, who were then distributed according to national needs. Most went into copper, lead, and zinc mines. Lack of experienced miners, meanwhile, hampered Colorado throughout the war.

Neither was government involvement limited to just one agency. The Metals Reserve Company, created in June 1940, purchased critical metals and minerals both at home and abroad, set ceiling prices, and paid premiums for production in excess of fixed quotas. The Defense Highway Act of 1941 provided for construction, maintenance, and improvement of access roads to sources of raw materials. From July 1942, until the end of the war, seventy roads were built in Colorado, most under ten miles, a few over twenty. The Strategic Materials Act provided for study of essential mineral sources. A Colorado field office opened in 1942 and was promptly swamped with requests to examine nearly five hundred mines. Last, but probably most important, the Federal Government sponsored scientific research (the atomic bomb comes immediately to mind), some of which directly affected Colorado mining.

The old-time prospector would have gaped in amazement to see Uncle Sam becoming virtually a silent partner in mining. Mining did continue in Colorado, though gold and silver production obviously dropped substantially. Boulder County, on the other hand, resumed its activity in producing tungsten, a war-sought material. Coal mining increased almost everywhere in the state. Moffat County's total jumped two and a half times between 1941 and its peak year 1944, when 138,000 tons were mined. Weld and Las Animas counties did not show such a jump, but these two major producers did reach the one-and-a-half-million-ton-per-year range and stayed there.

As the war drew to a close in 1945, L-280 was rescinded. This was no help to Colorado gold or silver mining, now at their lowest ebb since 1873. A shortage of manpower and the expense of getting back into production account for much of the decline, but interest in exclusive gold and silver mining stayed low for the rest of the forties. The two precious metals were mined but were generally by-products of base metal operations. The fixed price of gold, weighed against the high post-war costs of materials and rising wages, caused the operators to doubt the wisdom of continuing. Thanks to the dredge scouring the riverbed near Fairplay, placer gold production rebounded to a moderate level, the only segment of the once-dominant precious metals industry that held its own.

Such a dismal picture did not permeate all of Colorado mining, by any means. Zinc regained its earlier stature and maintained Eagle County's prominence as a mining area. From 1937 through 1948, the Red Cliff district (home of the Eagle Mine) led all state districts in the total value of copper, silver, lead, zinc, gold, and silver mined. Although lead and zinc had been somewhat scorned in the thirties, this was not true after 1941, and the Eagle Mine became the greatest zinc producer in Colorado's history. Also helpful were the government-paid subsidies for copper, lead, and zinc; then in 1947 controls on these metals were dropped, allowing price boosts to meet the rising mining costs. Gold miners were not so lucky.

Colorado suffered a blow to its pride, if nothing else, when in 1947, after twenty-three consecutive years as the premier American molybdenum state, it was surpassed by Utah. Recouping the honor temporarily the next year, Colorado was destined from then on to face competition from Utah, where molybdenum was recovered as a by-product of the copper operations. Climax Molybdenum continued to be the state's leading and sometimes only producer. It was an impressive, highly industrialized operation by the late forties. It gave the town of Climax such things as a large company-built auditorium for motion pictures, dances, or meetings. A brief recession in the market right after the war failed to affect molybdenum's future.

After years of waiting in the wings, northwestern Colorado oil fields came into the spotlight. Rangely,

141. Miners starting their shift at the Gilman Mine about 1947. Equipment had changed over the years, and electric power and lighting were now a must, as were hard hats and battery-operated lamps. Courtesy the New Jersey Zinc Company, Gilman.

142. The Rangely field came into prominence in 1946 with the completion of pipelines. By 1949, eighty-five percent of the state's oil was produced here—more than fifty thousand barrels a day. In a long isolated and forgotten section of Colorado, Rangely became the center of feverish activity. Courtesy the Denver Public Library, Western History Department.

143. A new and fearsome weapon, the atomic bomb, gave impetus to the Cold War. Suddenly Colorado's uranium reserves took on international significance. This was the mill at Uravan, near some of the important discoveries. Courtesy the Western Historical Collections, University of Colorado, Boulder.

huddled in the corner of the state, a mere "outpost in a desolate area," as one bemused viewer described it, launched the oil era. Early in the century the region was recognized as having possibilities for oil production; however, it was so isolated as to be of little commercial interest. In the 1930s and early 1940s, test drilling confirmed the earlier estimates, and in 1943 a producing well was brought in. The number crawled up to four the next year, leaped to thirty-three in 1945, before rocketing to 395 in 1947. Until 1945 transportation difficulties retarded development, crude oil having to be hauled by truck to either Craig, 107 miles distant, or Salt Lake City, 210 miles. Neither route provided particularly good roads, nor were the trucks large enough to carry much oil. This jam was broken in September 1945, when a pipeline from the field was tied into a Wyoming line serving Salt Lake City, Cheyenne, and Denver. Without pipelines to transport oil the fields could not reach their ultimate potential, the pipeline being to oil what the railroad had been to mining earlier. With the field prospering, the state got busy—a new highway from Craig cut the distance between there and Rangely down to 79 miles, helping measurably in the shipment of equipment and supplies. A second pipeline in 1948 increased production even more, and as the decade closed, Rangely was responsible for eighty-five percent of total Colorado oil production.

Less successful, but full of potential, were the large oil shale reserves of northwestern Colorado. Since the 1920s not much progress had been made in developing an economical and successful way to release the oil trapped in the shale. Wartime needs spurred the government to renewed efforts, and in 1944 an experimental program was authorized. Ten miles west of Rifle, on the Naval oil shale reserve, construction was started on an experimental mine and processing facility. The plant began test runs in 1947 and 1948, before being completed the next year. The Anvil Points project included offices, a laboratory, a refinery, a residential area, and shops. Some shale oil was converted into diesel oil (used by trucks and equipment on the site), and a few other products, including gasoline, were refined, but the process remained experimental. The cost of mining and refining simply could not compete with the low-cost oil pumped from Rangely and the other large fields.

Anvil Points, so named because of three nearby protruding cliffs that looked like blacksmith's anvils, very much resembled the nineteenth-century mining communities. The reaction of the people certainly was similar. One young lady remembered that, to her father, "It was a whole new field, a challenge," while to her mother, "it represented the ends of the earth. A place where only a totally insensitive person could survive." To the commentator it was an adventure, "a dream come true." A widely diverse group of people lived there and, to the amazement of the nearby homogeneous agricultural people, proved they could get along and work side by side. The site looked like some of the earlier jerry-built camps. The office plant was built from materials salvaged in Marble, Colorado, and houses included World War II vintage prefab housing and aluminum homes purchased in Denver. Well into the 1950s, Anvil Points had the bustling appearance once so familiar to mining communities.

An old-fashioned mining boom, with nearly all the trappings of earlier ones, also revived past glories. A variety of circumstances set the stage for it. During the war, the Manhattan Project had produced the atomic bomb, the terrifying result of which had been presented to the world in August 1945. Colorado thus assumed tremendous importance as the largest known source of uranium (Uravan and Central City districts), a prime ingredient of this new weapon. Finally, real peace had not come after the war; instead, the Cold War began, with the Russians and their allies on one side, and the United States and its allies on the other. To many Americans, it seemed that their only hope for survival rested with the country's atomic capability. Colorado held a trump card.

More and more uranium was needed to make bigger and bigger bombs, to conduct more tests, and to stockpile. The government was not about to leave the country's fate in private hands, or to allow the individualism which had characterized mining heretofore. These were the 1940s, and uranium was vital to national security. As a result, a government-nurtured rush was generated. Only the rewards were publicized; information that might give aid and comfort to the Communist bloc (quantities, locations) was carefully guarded. In 1946 the Atomic Energy Commission was established to take over development and production. The AEC formed a division of raw materials to organize exploration, procurement, and processing, and withdrew nearly 150 square miles of land from the public domain to be leased to uranium mining companies. A branch office was opened in Grand Junction, on its way to becoming the Colorado uranium capital. No person, unless licensed by the AEC, could transfer, deliver, receive, or export uranium; to insure milling capacity, the AEC purchased or rehabilitated mills in Utah and Colorado (Durango and Uravan). In 1948 a series of graduated price schedules was instituted to stimulate further the search for and mining of uranium. Government control was nearly complete in all branches of the industry.

Montrose County, where the U.S. Vanadium Corporation had been active for years with its mill at Naturita, captured most attention, although prospectors were becoming interested in all of Colorado's western border country. As far away as Gilpin County, and even in Boulder County, uranium fever rampaged. Old-timers' memories were jogged when *Look Magazine* proclaimed to the whole country in the December 21, 1948, issue, "Atomic Ore Discovered in Colorado." The long-forgotten Caribou Mine and surrounding district were suddenly jerked into the twentieth century. The 1950s promised to be an exciting time for uranium miners. Colorado could, in truth, look ahead with more hope for mining than she had been able to in decades.

144. The Caribou had been a major silver producer decades ago. In 1948 a vein of pitchblende was discovered, and the mine was jerked into the post-war era. It enjoyed brief nationwide publicity; however, the deposits proved small and within ten years the Caribou closed again. Courtesy the Colorado Mining Association, Denver.

11

Of Men and Machines—A Photographic Essay

Sweat and muscle open and operate a mine. Whether the miner labored in a hardrock or coal mine, or in a quarry, his day was dirty, noisy, and tiring. At first he had little but a strong back, wheelbarrow, hammer, drill, pick, and shovel to inch his excavation ahead. Soon blasting powder allowed him to loosen more of the mountain, and then, by building a track over which to push a car, it was easier to remove the rock. Hitching a mule to the car simplified the effort even more. A horse-operated whim or steam-operated hoist made it quicker and easier to transport the ore to the surface. Still, the insatiable quest for riches demanded more tonnage and faster work. A few miners evolved into crews of men, machine drills replaced hand drills, blasting powder became more powerful, electricity superseded steam power—and so on in the race to find the hidden riches.

Colorado pioneered several labor-saving devices. The first use of a power drill in mining in the United States occurred at Silver Plume in 1868–69 in the Burleigh tunnel. The water drill was invented by Coloradan John Leyner; it not only allowed faster drilling, but, more significantly, cut down rock dust, the scourge of mining. In milling and smelting, the Wilfley concentrating table, the Bolthoff-Boss pulverizer, and the "Colorado" stamp mill modernized processes. Mention has already been made of the early Colorado acceptance and use of electricity in mining. Other examples could be given; these, however, indicate the progressiveness of Coloradans in the field of mining.

The equipment of the modern mine hardly resembles that of a hundred years ago. Labor-saving machines and improved equipment have come to replace some of the manual labor, though not all. The story of a mine, however, remains its miners. Working deep underground, they still blast, shovel, and haul, now for a higher wage per hour than their nineteenth-century counterparts received in several days.

145. Colorado mines came in all shapes and sizes, the majority returning little or no profit to optimistic stockholders and owners. It has been said that more money went into the earth than was ever taken out in ore. For every bonanza there were scores of busts. This old-timer working in Saguache County probably made enough to provide himself with a living. Note the candleholder, commonly used before lamps. Courtesy the Denver Public Library, Western History Department.

146. This determined crew was just getting started. Here the boss looks over the ore. The two miners in the background are double jacking, a popular drilling method. Courtesy the United States Forest Service, Washington.

147. The Gold Coin Mine illustrates what all owners hoped their property would become. According to story, the vein was found during excavation of the site for a Victor hotel. Courtesy the Cripple Creek District Museum, Inc., Cripple Creek.

148. Leadville's AY and Minnie gave the Guggenheims a fortune before they turned to smelting. Note the log cribbing retaining the tailings and dump, a familiar sight in Colorado mining regions. Courtesy the United States Geological Survey, Denver and Washington.

149. The underground miner dressed as he pleased, his attire depending to a great degree on whether it was a wet or dry operation. In this 1909 photograph, candles were still being used, as the men double jacked their powder hole. Courtesy the United States Forest Service, Washington.

150. High explosives eased the miners' work considerably. This man was setting his fuses at the breast of a drift. The fuses burned at a specified rate and were set to go off at different intervals to achieve the maximum breakage. Many a miner was injured when a charge failed to fire and he mined into it. A careful count was attempted to see if all charges exploded according to plan. Courtesy the Center of Southwest Studies, Fort Lewis College, Durango.

151. A great help in transporting rock was the ore car, commonly the one-ton car. Moving on an easily constructed track, they could be used singly or linked together and pulled by a mule or horse. Vanadium was being mined in the Primos Chemical Company mine in San Miguel County. Courtesy Mrs. Homer Reid, Telluride.

152. Work inside and outside the mine required muscle to keep the operation going. This crew of hearties was hoisting the powder house up to a higher level of the Virginius Mine. The Virginius was lofty to start with, being situated at 12,000 feet. Courtesy the State Historical Society of Colorado, Denver.

153. Animal power provided the obvious initial solution to power needs; thus, horses, mules, and burros were drafted. This horse-power whim, an easy adaptation, worked simply: a change in the animal's direction raised or lowered the bucket. Courtesy the State Historical Society of Colorado, Denver.

154. Mules pulled ore cars for nearly a century in Colorado, as this 1950 photograph, taken in the CF&I's Morley Mine, shows. These animals were the mainstay of mining, working long hours under trying conditions. Courtesy the Colorado Fuel & Iron Steel Corporation Archives, Pueblo.

155. In developed mines the cage replaced the bucket. Men, materials, and cars could be easily lifted to the surface. One of the high injury zones of mining was the shaft, where men were hit by falling objects; hence the roof over the cage. Eventually the safety brake (to insure against unscheduled falls) and enclosing mesh on the sides produced the safety cage. This was the 1896 crew at the Victor Mine, Teller County. Courtesy the Denver Public Library, Western History Department.

156. The engine and hoist room were the heart of the mining operation. The miners' lives depended on the skill and knowledge of the hoist operator. By the use of the wheel gauges, he knew exactly at which level his cages were; the standard bell signals told him to raise or lower them, or that an accident had occurred. Courtesy the State Historical Society of Colorado, Denver.

157. The power drill revolutionized mining in the 1880s, and the cost was not great. Initially called a "widow maker" because of the dust-related deaths caused by razor-sharp silica dust, the drill was improved by water shooting through the hollow drill steel, thus flushing out the hole, cutting dust, and also cooling the bit. The man in the background is single jacking. Courtesy the United States Forest Service, Washington.

158. A drilling team in Winfield Stratton's Independence Mine. Courtesy the State Historical Society of Colorado, Denver.

159. Over the years the design of the drill changed, but its functions remained constant—to cut the work and time needed to drill a hole. These men were plugging with a stopehammer in one of the U.S. Vanadium mines. Courtesy the Colorado Mining Association, Denver.

160. The power of the drill could be used for many jobs. One of the dangers of mining, particularly of coal, was having the roof cave in or a slab fall. These men are rock bolting the roof to stabilize it and increase safety. Courtesy the Colorado Fuel & Iron Steel Corporation Archives, Pueblo.

161. The compressor was as necessary to mining as the pick, once the power drill was accepted. It was used for power, ventilation, and cooling air in hot mines. Permanent compressors, or a portable gasoline model such as this one near Ouray in 1935, were found everywhere. Courtesy the Western Colorado Power Collection, Center of Southwest Studies, Durango.

162. It was foredoomed that mechanized hauling would replace the horse and mule. This muck train illustrates the change that had come by the mid 1950s. More tonnage could be hauled at a lower cost. Courtesy the Colorado Mining Association, Denver.

163. Soon specially built cars were transporting men to work with comfort and speed undreamed of in earlier times. This electric-powered man trip car was luxurious for the mid 1950s. Courtesy the Colorado Fuel & Iron Steel Corporation Archives, Pueblo.

164. Where railroads could not go, and pack trains only at great expense, the tramway was the answer. Curving around mountains, from one tower to the next, it provided an economical means of transporting ore out and bringing supplies and occasionally a brave rider in. Gravity flow, full buckets going down pulling up empty ones, helped provide motive power. Colorado pioneered acceptance of trams, and the San Juan region especially took them to heart. Courtesy the State Historical Society of Colorado, Denver.

165. If improvements were taking place within the mine, so were they in the placer fields. Hydraulicking was imported from California to handle low-grade operations at a reasonable cost. This nozzle shot a stream of water four hundred feet at the Keystone placer near Telluride. Unfortunately, the returns here proved poor and the mess great, problems which eventually killed hydraulicking. Courtesy the United States Forest Service, Washington.

166. The Colorado placer miner could be as inventive and ingenious as his counterpart in other segments of American industry. Operating his machine in north Clear Creek, this man made between $1.50 and $2.50 per day in 1915. Black Hawk is in the background. Courtesy the United States Forest Service, Washington.

167. A steam shovel and a rotary screen, for separating pay dirt from rock, were part of the San Juan Metals operation in 1937. Mining had come a long way since pan and shovel days. Courtesy the Western Colorado Power Collection, Center of Southwest Studies, Durango.

168. Meanwhile, underground the hardrock and coal miners found machines aiding them in a variety of ways. Timber trucks on the left and ore cars on the right carried more loads faster than had their predecessors. Courtesy the New Jersey Zinc Company, Gilman.

169. Nor did the miner have to dump his load by hand anymore. This revolving dump car did the work for him. Courtesy the Denver Public Library, Western History Department.

170. Even though oil shale had not proved profitable, the mine was modernized. This operation near Anvil Points indicated that the tonnage could be mined, if the refining problem could be overcome. Courtesy the United States Geological Survey, Denver and Washington.

171. There was even a machine to cut coal. This short wall coal cutter operated near Axial, Moffat County, in 1960. Courtesy the United States Geological Survey, Denver and Washington.

172. The coal loader and conveyor belt simplified the removal of coal from the mine. The cost of equipment such as this was a major reason that mining became the business of large corporations with plenty of financial reserves. Courtesy the Colorado Fuel & Iron Steel Corporation Archives, Pueblo.

173. A corporation such as CF&I could afford to modernize, thereby increasing production per man-hour. This "continuous miner" was introduced into their Allen coal mine in 1967. The noise and dust were not reduced, but more coal was produced. Courtesy the Colorado Fuel & Iron Steel Corporation Archives, Pueblo.

174. Safety was a factor that concerned everyone involved in mining. This machine was installed in the Delagua coal mine about 1912 to mix adobe or rock dust with coal dust, thus helping to prevent coal dust explosions. This method, with improved equipment, would be used for decades. From the author's private collection.

175. Preventing cave-ins was a constant battle. This longwall equipment of the 1970s shows how far the effort was being carried. Courtesy the Colorado Fuel & Iron Steel Corporation Archives, Pueblo.

176. To the assay office the prospector took his sample and the miner his ore; the company assayer was crucial to the mining operation. Field assaying, such as this, could seal a district's fate. The man to the left was using the familiar mortar and pestle to crush the sample, and the fellow on the right seems to be studying a sample with a magnifying glass. Courtesy the State Historical Society of Colorado, Denver.

177. The modern mining company needs a well-equipped chemistry lab and assay office, a far cry from what appeared in the previous 1897 photograph. Assayers were not always popular—or honest. A great deal hinged on their reports and one man wrote that he thought the local assayer "made the ores yield by certificates more than the truth." Courtesy Mrs. Homer Reid, Telluride.

178. One of the first reduction methods tried in Colorado was stamps to crush gold ore. By the turn of the century, stamps were found in multibatteries, as only one phase of the total process. Smuggler Union mill at Telluride. Courtesy the State Historical Society of Colorado, Denver.

179. Mills became more complicated and recovery percentages higher. From left to right in the Gilman mill's grinding department were found a rod mill, Akins classifier, ball mill, and another classifier, after which the machinery repeats itself. Courtesy the New Jersey Zinc Company, Gilman.

180. One of Colorado's important contributions was the Wilfley table. As finely crushed ores flowed over the agitating table, the different specific gravity of each mineral deposited it on a predetermined bar. The table proved particularly valuable in separating zinc and spurred interest in that metal. Courtesy the Boulder Historical Society, Boulder.

181. The smelter crew of the Caribou Consolidated Mining Company in Nederland in the early 1880s. Courtesy First Federal Savings, Denver.

182. Without question one of the finest dressed mill oilers at the New Jersey Zinc's Gilman operation. Courtesy New Jersey Zinc Company, Gilman.

183. Initially machinery came primarily from Chicago; eventually new or improved machinery and equipment could be manufactured right in the state, a notable factor in the growth of the mining industry. Over the years, Colorado had many companies, of which the Colorado Iron Works is representative. Courtesy the State Historical Society of Colorado, Denver.

184. Electricity generated the power and light in Colorado mines after the turn of the century. This was the Ames station, which generated 3,000 volts of alternating current in 1891 for the Gold King Mine over 2.6 miles away, a first in the electrical industry. By 1905 the original building had been joined by others. Courtesy the Western Colorado Power Collection, Center of Southwest Studies, Durango.

185. Logging was and is an important subsidiary industry of mining. Timbers shored the mines, hungry engines consumed cords of wood, logs cribbed the dumps and tailings piles, and lumber went into homes and buildings. These logs were being hauled to the Virginius Mine. Courtesy Ruth Gregory, Ouray.

186. The mines of Colorado did not swallow as much timber as some other mining areas, particularly the geologically unstable Comstock, but they consumed their share. An example of how timbers were used, in an unidentified mine. Courtesy the United States Forest Service, Washington.

187. Inside the mines, timbering was elaborate if need be. The diamond stope of the Portland Mine, with square-set timbering made famous by the Comstock, was a classic example. From the man standing at the bottom, an idea of the size of the stopes can be gained. Courtesy the United States Geological Survey, Denver and Washington.

12

One Hundred Years and Going Strong

The promise of the 1940s was fulfilled during the next twenty years, and Colorado regained its preeminence in the field of mining. Uranium and oil, particularly, caught both the public's and the miners' attention.

The uranium boom, fostered by the AEC, paid rich dividends in the 1950s. Suddenly, land in Montrose, San Miguel, and Mesa counties, which had never been considered to be more than marginal acreage suitable for ranching, became the center of frenzied activity. Gone were the prospectors, with their noses for ore, and trusty burros. They were replaced by men carrying geiger counters, riding in jeeps—or airplanes— and using scintillation counters to pick up traces of uranium. There were inspiring success stories, those of Vernon Pick and Charles Steen over in Utah being the best known. And there were booklets to tell the weekend amateur how to find uranium. One pamphlet subtitled a chapter on where to find uranium deposits "Or Try the Crystal Ball." No doubt, some used bizarre methods, but more reliance was placed on scientific instruments and professional education than ever before. Like the guidebooks of 1859, these modern counterparts told what to wear ("tough, relatively heavy pants and shirts"), what to take (an air mattress "may be found very serviceable"), and encouraged one and all ("uranium has become a magic word for many Americans in the past few years").

And a certain kind of magic it was. Grand Junction became the goal for mining engineers, promoters, miners, government employees, and the curious. The profusion of long distance calls forced the harassed phone company nearly to double its number of circuits between 1952 and 1954 in order to handle the business. Uranium stocks became the "new," quick way to wealth, and the sucker again fell prey to the wily promoter who offered shares in companies formed even before core drill tests had been made. As far away as Denver, a well-known dry goods store set up a "Prospectors' Corner," with books and necessities for hunting uranium. Few merchants on the western border failed to stock merchandise that would tempt the twentieth-century prospector.

Uncle Sam kept a watchful eye upon it all; not until 1956–57 did the government relax restrictions and allow tonnage production to be announced. Price supports and other assistance continued, and uranium production responded predictably. By 1955, 174 producing uranium mines, many small-scale, were scattered about this land of mesas, canyons, and sand. The pitchblende deposits of eastern Colorado never were much of a factor, despite great expectations and no little newspaper promotion. To process the ores, plants were operating at Rifle, Uravan, Slick Rock, Maybell, Durango, Naturita, Gunnison, Grand

188. The Anvil Points oil retort had come a long way from its primitive ancestor of the 1920s; however, it was still too costly to recover oil from the shale. The industry went through ups and downs in the 1950s and 1960s. Courtesy the United States Geological Survey, Denver and Washington.

Junction, and Canon City. The Naturita Mill was closed in 1957; as the excitement waned, others followed suit.

The magic faded in the 1960s, mainly because the AEC began to curtail its support and limit its purchases. The first full year of reduced federal support, 1962, found the total value of uranium down $12.1 million. At the end of 1969, uranium production for the year was barely a third of what it had been ten years before. The uranium boom was over; a degree of quiet returned to the plateau country, scarred now by dumps, roads going nowhere, and relics left by the twentieth-century prospector and miner.

Production of vanadium stayed fairly constant and in 1969 was a little over what it had been a decade earlier. When the uranium boom passed, Colorado saw one more mining excitement fade away, perhaps the last in which an individual had a chance to make a fortune by a lucky discovery or sale. Mining had become big business, and money and technology spoke louder than individual initiative or luck. The uranium era brought a measure of prosperity and increased business to western Colorado—to those directly involved, to subsidiary mining operations (such as the Rico Argentine Mining Company in Dolores County, which produced sulphuric acid needed by the processing mills), and to a variety of auxiliary workers, from ranchers to carpenters to mobile home salesmen. It also helped popularize the area and, to a degree, changed the topography. Working through the Bureau of Public Roads, the AEC built access roads throughout the region, opening it to the public as never before.

Less fascinating to the public but producing more wealth was the Colorado oil boom, which paralleled that of uranium. The spectacular success of the Rangely field in the forties undoubtedly created interest in searching for other major deposits. In the early fifties the Denver-Julesburg basin field was opened, a worthy competitor to Rangely. The old Boulder oil field, known for years, was at the western end of this basin, but the scope of it was not appreciated (though discoveries were made in Weld County in the 1930s) until oil was found in neighboring Nebraska in the summer of 1949. This set off a wave of surveying, seismic testing, leasing, and drilling which launched an oil boom in northeastern Colorado. Rangely and the Denver-Julesburg area were the twin pillars which supported Colorado as the number two oil state in the Rocky Mountains, behind Wyoming. Interest spread, and the state's southwestern corner also came into prominence, led by La Plata County which produced both oil and natural gas.

Crude petroleum production rose steadily until it reached a peak of fifty-eight million barrels in 1956. By mid-decade seven petroleum refineries were operating, five in Denver and one each in Alamosa and Rangely. Pipeline planning and construction kept pace, as did exploration and wildcat drilling. Oil wells proved fickle, however; the business was one of high odds. Of the 1,539 wells drilled for oil and gas in 1955 (a peak year), 1,043 were dry. This was not unusual. Sixty-six percent of the 99 wells drilled in 1950 were dry, slightly less than the seventy-three percent in 1958. These lengthy odds failed to quash the excitement endemic to an oil boom such as this, and the bigger companies used all the technological skill they could muster to shorten them.

Counties which had known little or no mineral activity suddenly were swept into the midst of the boom. The farmers and ranchers of Logan and Morgan counties watched the oil man elbow his way onto the scene. Wide, unbroken high plains vistas were now framed by derricks and storage tanks. Conversations in Fort Morgan and Sterling revolved around the price of oil, rather than around cattle or sugar beets. The change was noticeable everywhere. In Morgan County the population jumped twenty-two percent in seven years, and the assessed property valuation rose two and a half times during the same period. New industries, such as natural gas refining, came with the oil, and local merchants found their businesses thriving. Profits lured competitors so it did not prove an unmixed blessing. The same conditions prevailed in Logan County, which was the second leading oil producer in 1956. After that it settled down to steady production. La Plata County, which had been through mining rushes earlier, had never before experienced such a period of economic growth. The assessed property valuation jumped three hundred and fifty percent. Durango became the oil companies' headquarters and prospered and grew as it had not in decades.

Another evidence of the changing times was the response of the state government. The first petroleum law, adopted in 1889, prohibited emptying oil into any creek, stream, river, or lake. Scarcity of legislation during the next several decades indicated the tardy development of the industry. Not until 1927 was a regulatory body, the Gas Conservation Commission, established; two years later legislation was passed to regulate well plugging and abandonment methods. Then during the next twenty-two years no major changes were made. The sudden emergence of oil in the postwar years spurred the legislature into action, and in 1951 an oil and gas conservation commission, with a staff of twelve, replaced the older body. Rules and regulations governing production and waste appeared; in 1953 so did a state severance tax on oil production.

It was not the new fields alone that grabbed the headlines. The Florence field continued to produce on a small scale, and Well No. 42 did more than its share. Drilled in 1889, it was still yielding a few barrels every two or three days in 1959, having produced by then an estimated million barrels of oil.

A decline was noticeable in Colorado's crude oil production as early as 1957, evidence of the stabilization of the Rangely field and the waning of several of the Denver-Julesburg basin districts. Oil production continued its gradual decline in subsequent years, although the worth of its yearly production put it far ahead of most mining industries throughout the 1960s. Only the Rangely field maintained its production near earlier levels, which by 1969 boosted Rio Blanco County into the number one position in yearly and cumulative totals. Over half of the state's production that year came from that county, with Washington and Logan counties in the Denver-Julesburg basin trailing far behind in second and third places.

During these two decades hope persisted that oil shale would finally bestow some profits upon those who worked with it. The high expectations for the governmental plant at Anvil Points were dashed in 1956–57, when Congress suspended tests and put the site on a stand-by basis. Over 500,000 tons of shale had been mined and tested, and the conclusion was reached that commercial production of these deposits must wait until mining and processing costs were reduced to a level commensurate with that of finding and producing ordinary oil. The rather discouraging possibility (for the consumer) was held out that, as the expense of finding oil went up, the cost differential would be reduced and might provide the needed impetus for shale. Experimentation went on. Union Oil opened a pilot plant near Grand Valley; it soon closed, however, and was dismantled early in the sixties. The costs remained too high for both the government and private companies.

Amazingly enough, hope never evaporated. In August 1960, the AEC set off a thousand-pound charge of explosives in a Rifle mine to test the feasibility of

189. A long cherished dream of Colorado silver miners was a higher price per ounce for silver, a price free from federal regulation. When it came, the results were not as great as expected, but some mining started up, such as the Homestake's Bull Dog operation near Creede. The mill was being constructed with machines that would have dazzled the nineteenth-century builder. Courtesy the Homestake Mining Company, Lead, South Dakota.

releasing oil directly by using nuclear explosives. It was a good try but yielded little. Then, in May 1964, the government leased its Anvil Points facilities to a combine of oil companies for a five-year study. Activity by decade's end had come to a standstill; the economic logjam had not been broken.

Less exotic, but immeasurably more successful, was the old standby, molybdenum, which maintained its position during these years as one of the top-value minerals in Colorado. Climax held its rank as the dominant producer, and the company gained control of the one other significant mine, the Urad in Clear Creek County, thereby cornering production. Even a six-month strike in 1962–63 only temporarily slowed operations. In May 1964, Climax detonated 417,216 pounds of explosives to break loose a million and a half tons of molybdenum. Reportedly the largest blast ever set off in United States mining history, it jarred seismographs as far away as Yellow Knife, in Canada's Northwest Territory. Within hours after the explosion, the molybdenum ore was into production.

In 1968 the Climax Mine, the world's largest single source of the metal, celebrated its fiftieth anniversary. In 1918 the mill could handle 250 tons per day; by 1968, it was handling 43,000 tons. In 1969, the one hundred millionth ton of ore was taken out. The company employed some 2,200 people; many, in support positions, had never seen the inside of the mine. Production for the year 1969 totaled over $90 million (the rest of the state's molybdenum came from the Urad Mine), a gigantic, even meaningless figure, until one realizes that this was more than the total value of any other mineral or mineral fuel in Colorado, and was more than the combined total of the next ten ranking minerals.

Molybdenum was not all that the Climax company was mining; as by-products, tungsten, pyrite, and tin concentrates were being recovered, along with several lesser minerals. Climax and Boulder County had accounted for well over ninety-five percent of the state's total tungsten production since the first shovelful had been turned. In 1969 Climax produced nearly all the tungsten and all of the pyrite and tin in Colorado. This one company, with two mines, represented more than half of the total value of Colorado's mineral production that year.

The Eagle Mine, though never able to dominate zinc production so completely, remained the largest zinc producer and retained its position as one of the important lead and silver mines. It suffered serious

190. An era ended when this dredge near Fairplay finally quit operating. It left behind its pond and a raped riverbed, paralleled by mounds of gravel and rocks. Courtesy the United States Geological Survey, Denver and Washington.

labor troubles in 1952 and again in 1960 but was hurt more by a fluctuating zinc and lead price. After high levels in the early fifties, zinc production leveled off, then picked up again in the sixties, before slumping again late in the decade. San Miguel County, basically the Idarado mining operations, followed Eagle County in production.

San Miguel County, meanwhile, emerged as the leading producer of silver, gold, copper, and lead, and was second in zinc, uranium, and vanadium. The Idarado was its largest facility and led the state for all base metals; gold and silver were by-products. The San Juan mining region continued to show its mineral diversity and was second only to Lake County in mineral importance. The San Juans were making a strong comeback from the depressed years of the twenties and thirties.

That old stalwart—gold mining—never regained its earlier dominance and continued to slump. Cripple Creek, long the prime district, had the advantage of the brand new Carlton custom mill, completed in the early fifties, with its latest labor-saving devices and improved equipment. Teller County consequently led in gold production throughout the 1950s. One significant figure indicated the statewide trend: in 1958, forty-three percent of the gold and silver recovered in the state came as by-products of base metal mining. An era closed three years later, when mining at Cripple Creek ceased for the first time since 1891, and only clean-up operations at the Carlton Mill continued into 1962. The foremost gold producer in Colorado, and the second largest gold district in the United States, had run dry. This district had produced precious metals exclusively, and it could not change as the San Juans had done on several occasions. The Colorado Bureau of Mines report of 1969 listed only sand, gravel, and peat production for the district.

Another era ended when the dredge at Fairplay, the last major placer gold operation, shut down in 1952. Silver did somewhat better, basically because its price, freed in the early sixties from government regulation, rose to $1.75 per ounce, the highest in Colorado's history. Old-timers had been fond of saying that $1.30 for silver would bring back the great days again. But only sightseers swamped the old silver districts.

The reasons for the decline of gold mining and the failure of silver to revive were varied. Miners' wages rose steadily, as did the costs of supplies and equipment. In most areas the high-grade ore had long since been taken out and only lower-grade ore remained, leaving the miner in the age-old bind of higher costs vs. declining ore value. Lack of smelters hurt silver mining, and the inflexibility of the gold price squelched enthusiasm. Less significant factors, such as mounting state and federal mining regulations, a shortage of skilled miners, and concern over environmental issues plagued the operators. It was costly, if not prohibitive, for the poorly financed individual to try to reopen an older mine, or continue operations, unless he found good ore. The big companies, already in business, profited when silver went up in price, as did, to a lesser degree, the smaller operator who was in production; to start fresh, however, was a formidable task.

Coal was faced with many of the same problems as gold, plus a few more, some of which threatened the very existence of the industry. As the natural gas and oil fields opened and expanded, industry and private homeowners switched to these cleaner and more efficient fuels. In the 1950s, times were hard for the coal miner, who was confronted by a marked decline in coal usage. Only in the sixties was the situation reversed. Finally, in 1964, production neared the 1949 level. The number of operating mines diminished

191. The Climax Company continued to dominate molybdenum. In 1957 it merged with American Metal Company to become AMAX-American Metal Climax, Inc. The Urad Mill near Georgetown, a long-time producer, closed in 1974. A sign of the changing times was the company's willingness to restore the site through a broad conservation program. Courtesy the Colorado Mining Association, Denver.

alarmingly, from 150 in 1955, to 82 in 1964, to 54 in 1968.

Sometime in 1957, an unknown miner (underground or perhaps operating a machine at a strip mine) took out the 500 millionth ton of coal mined since the 1860s. It was ironic that this milestone should be passed during such a depressed time for coal mining.

Coal strip mining became more prominent. The number of mines was never very high, but remained fairly constant. In 1955, there were seven; in 1968, five. Much of the strip-mined coal was used in electric generating plants; indeed, home coal use, except perhaps through electric power and lights, dwindled to nothing. Of the coal mined in 1968, fifty-seven percent went to power plants and thirty-nine percent to steel mills. The coal stove and coal furnace, except in isolated instances, were things of the past, as were the local coal company and the dirty coal bin in the basement of the home. There was still a bountiful supply of coal reserves in Colorado, but the demand and the market were not there. A few prophets warned that some day these deposits would be vitally needed, but at that moment they were ignored.

An overview of these twenty years finds the increase in mineral production, which started with the end of World War II, continuing in the 1950s but declining in the 1960s. Various minerals and mineral fuels rose or fell independently within this general pattern, as previous comments have indicated. After a flurry of activity in uranium mining, direct government involvement dropped off, but not government regulation. Regulations, reports, rules, and "red tape" mounted. One of the significant postwar trends in mining was the continuation, in one form or another, of Uncle Sam's surveillance. Ever since the thirties, this trend had been growing and it received a boost when mining became a strategic factor in the Cold War.

Another 1960s trend was the growing cost of all aspects of mining, which threatened to force many small operators out of the industry. This was a reflection of the increases generally in the American cost of living, and of the miners' demands for higher wages and better working conditions. Even with higher wages, it was becoming harder to find experienced miners. Nor could the mechanized mine be opened or maintained as cheaply as the single-jacked, blasted, shoveled, and animal-hauled ore operation of seventy years before.

As mining became more technical, the skill, training, and experience required to run it became more complex. Mining engineers and geologists were consulted on where to blast or where to drill. The cost was too high to risk a "by guess and by golly" type of operation any more; even with such expertise, the odds were long in the oil fields and only slightly better in hardrock mining.

Large companies increased their hold on Colorado mining; Climax Molybdenum was just the twentieth-century evolution of a Tomboy Mines Limited or a Little Pittsburg Consolidated. Except during the early uranium excitement, production of oil and the base and precious metals was dominated by corporations, whose headquarters were elsewhere and which were run for the benefit of stockholders who never visited a mine, let alone worked one. Their Colorado mine holdings were commonly only a small part of a conglomerate whose interests were worldwide.

The old mining camps, those that had not slipped into the final silent sleep of a ghost town, found increasing revenue could be gained from tourists, who were engulfing Colorado in greater numbers each year (also benefiting the oil companies' profits). They

192. Strip mining and its ugly scars appeared in Colorado on a large scale. Underground operations had been carried on for many years at Oak Creek in Routt County; now lower-grade coal deposits fell to the strip miners. Courtesy the United States Geological Survey, Denver and Washington.

193. These are not visitors from another planet, but rather a miner rescue training class. Despite all the advances in mining safety and legislation, the threat of mining accidents still haunted the miners. Trained crews and new equipment promised at least to reach the site faster than in the old days. Courtesy the Colorado Division of Mines, Denver.

climbed over mine dumps, scampered through old buildings, visited newly opened souvenir stores, took rolls of pictures, generally poked around, and even listened to old-timers reminisce about the days of yore. Gold and silver camps attracted the most attention; no oil or uranium hamlet generated much interest. The fascination with the precious metals had not really died since 1859. Some camps, like Cripple Creek, seized the opportunity with jeep tours and a melodrama —only feebly replacing the lost mining economy. Central City went the furthest by turning itself into a tourist attraction, with everything from summer opera to noisy saloons and tourist traps. For a dollar or so, one could pan for gold along the stream banks, sensing for a moment the enchantment of a lost era—not the drudgery, disappointment, and hard work. Ghost towns became the goal for four-wheel drivers and resulted in more vandalism in the 1950s and 1960s than had occurred during the previous five decades; souvenirs were dug, ripped, or chopped out wherever they were found. The old buildings and abandoned sites had few protectors from the horde which invaded them. As had their nineteenth-century counterparts, the twentieth-century American littered just as thoughtlessly, and a multitude of pop bottles, cans, and refuse despoiled the site. Man and nature combined their efforts to erase what once had been alive and vital.

The years 1930 to 1969 saw Colorado's mining become depressed, then recover, only to decline again.

194. Old age was not kind to the camps. Some hung onto their existence by selling their heritage. Silverton, pictured here, was fortunate to have a narrow-gauge railway still operating, and it brought tourists in who kept the town alive until mining revived in the late 1960s. Courtesy the United States Forest Service, Washington.

Newly profitable minerals or mineral fuels injected fresh interest into the industry, which had stagnated after World War I. The glamorous precious metals maintained their fascination for the tourist and the average Coloradan, but the real money was to be made in the less interesting base metals, petroleum, and the briefly exciting uranium. Molybdenum and zinc dominated most of the mineral production, challenged for a while by uranium and vanadium. Coal had lost out to petroleum on many fronts and, while sustaining production, found its markets becoming limited to industry and electric power. Though the growth of urbanization along the eastern slope promised more industry and more consumption of mineral products, it also created mountain urbanization in mining areas. The sixties also heard more outspoken concern for the environment than had been expressed before, and talk of strict controls and restrictions on mining bespoke troubles ahead.

On September 10, 1969, in the Rulison area near Grand Valley, an atomic charge was detonated to determine its feasibility for freeing natural gas. Colorado had truly arrived in the atomic age with this first atomic blast within the state. Mining also was entering a new era, and the seventies and the last third of the century promised to be exciting and challenging.

TABLE 3
Production of Metals, Nonmetallics, Petroleum and Coal
State Totals

	1962 Production	Cumulative Through 1962
Metallics		
Molybdenum	$ 46,695,020	$ 613,032,788
Uranium	45,551,080	781,688,804
Vanadium	9,964,400	80,414,260
Zinc	10,346,567	377,599,139
Tungsten	3,052,781	41,281,453
Lead	3,270,316	333,610,658
Copper	2,530,305	101,655,543
Gold	1,719,102	915,719,529
Silver	2,123,519	595,175,508
Pyrite	836,480)	
Iron	170,789)	7,307,363
Beryl	178,316	1,130,289
Tin	61,348	436,783
Miscellaneous Metallics	342,850	11,190,120
Total Metallic Mineral Production	$126,842,873	$3,860,242,237
Nonmetallics		
Cement	$ 18,245,522	$ 143,676,206
Sand & Gravel	18,539,971	131,813,150
Stone	2,609,340	23,780,429
Limestone	1,926,382	33,394,754
Clay	1,612,010	17,082,993
Fluorspar	1,619,460	30,876,757
Feldspar, Mica, Gypsum	434,414	10,166,207
Volcanic Scoria, Pumicite	81,734	3,479,422
Peat Moss	72,944	276,710
Gemstones	40,000	272,678
Miscellaneous Nonmetallics	1,490,042	3,880,017
Total Nonmetallic Mineral Production	$ 46,671,819	$ 398,699,323
Mineral Fuels		
Crude Oil	$122,334,166	$1,767,230,853
Gas Sold	9,099,431)	
Gas (All classes, well head value)	11,688,103)	147,902,364
Coal	20,898,231	2,106,885,856
Total Mineral Fuel Production	$164,019,931	$4,022,019,073
TOTAL STATE MINERAL PRODUCTION	$337,534,623	$8,280,960,633

Source: Colorado Division of Mines

13

Challenge of the Seventies

Colorado mining faced new challenges in the 1970s and found that the old ones would not die. Both hope and disappointment surfaced, depending on the issue and the time. A miner of the 1870s would have recognized the symptoms, if not some of the issues.

The question most in the public eye as the decade unfolded was the energy shortage, evidences of which were steeply rising oil prices, a shortage of gasoline, and fears of rapidly falling oil and gas reserves. All these things spurred the search for new sources—oil shale, strip-mined coal, and natural gas and oil deposits. Both the government and private industry displayed great interest in the question, and rumors of ambitious plans and operations emerged.

The Rulison project in 1969–70 and the Rio Blanco project in 1973 both attempted to utilize nuclear energy to release natural gas but raised little except howls of protest. The Rio Blanco blast exploded three special bombs to tap a deep-lying natural gas formation which failed to produce as expected; unfortunately, radioactive traces appeared in some of the gas tested. This experiment generated protests from residents of the town of Rio Blanco, as well as from national environmental groups, about damage to private property, radiation, environmental pollution, and water contamination. The AEC intended to launch an era of exploration and mining based upon underground nuclear blasting, only to find itself under great pressure to cease. Evidence of the concern was a citizen-initiated law passed in the 1974 general election, which required a statewide vote to approve future underground nuclear blasts.

In the summer of 1976 the Rio Blanco and Rulison wells were plugged with cement, the ground seeded and graded, and the sites abandoned. Technically the Rulison test had succeeded, economically it failed. Rio Blanco failed totally, later tests showing that engineers had overestimated the amount of gas in the rocks that had been fractured by the blasts. Thus ended one of the most controversial programs in Colorado mining history.

As the energy crunch worsened, oil shale fever mounted, duplicating earlier episodes. Both Rio Blanco and Garfield counties were known to have large reserves. As early as 1971, reports predicted "unprecedented" growth for these two counties, as well as for neighboring Mesa County, when the oil shale fields opened. There was talk of population doubling, even tripling, with corresponding expansions of business, construction, and community services. What once would have been the delight of chambers of commerce now produced worries about environmental impact, excessive demand on limited water resources, disruption of local life, and the aftermath of depression when the boom passed. State and federal governments became involved and reports proliferated.

The Anvil Points facility was leased to a firm backed by a consortium of energy and oil companies, and a new plant, to cost more than $250 million, was planned northwest of Rifle. Little transpired, except that Anvil Points produced some crude oil for tests to determine possible uses of shale oil. The old bugaboo of finding a commercially competitive means to unlock oil shale deposits was partially to blame. So were uncertainties about federal energy policy, tight money during the mid-seventies, recessions, and inflation. What was once planned at a cost of a quarter of a million dollars, cost nearly a half. The continual pressure of environmentalists could not be ignored either. Thus Colorado's latest oil shale boom was deflated or, at best, postponed until better times.

An article in the July 4, 1976, *Denver Post* showed what had happened. Gulf Oil Corporation and Standard Oil Company asked the federal government to allow them to suspend operations for two years on 5,000 acres of leased oil-shale land and suspend lease payments. Environmental and development uncertainties stopped them, despite their having already made payments of more than $126 million. Other companies leasing oil-shale land were working at a reduced level or had pulled out, citing the lack of federally guaranteed loans for oil-shale plants.

Environmentalists were having a say in mining unlike anything previously heard of by Colorado miners. Concern about the environment had even taken hold of mining companies, but whether this reflected a sincere concern or merely public pressure was hard to determine. As they closed the Urad operation in 1974, the Climax people promptly launched a program to revegetate and landscape the tailings ponds and general site. The program, designed to take several years, if not decades, would not only conceal scars but stabilize the area to prevent erosion and water pollution. The project's scope was exciting—and costly. Never again would miners have the free hand they once enjoyed, but their predecessors should not be blamed for failing to undertake similar restorative procedures. It is not historically fair to measure them against standards only now being seriously implemented. There was environmental concern as early as the 1870s, particularly over water, but it lacked a broad appeal, a sense of urgency, or a total commitment.

As the miners were being watched in this area, they also came under increased government and state regulation, which either hampered or monitored their operations; the interpretation of their effect depended upon one's viewpoint. If the sixties had been regulated, the seventies were even more so. Regulation became a fact of mining life. Almost every time the operator turned around there was some new form to fill out or rule to be followed. Among the regulations were those of the Forest Service, which had that lovable bruin, Smokey Bear, looking over the miners' shoulders. An environmental analysis had to be conducted for all larger operations in national forests and a bond posted to insure the reclamation of the land. Regulations about the types of roads and transportation to be utilized were other examples.

The federal government seemed to have the ability to instigate the more vexing regulations. In 1973–74 it pushed for a second opening for every working area within the mine, which meant driving raises to the surface. Fortunately, after much protest, the order was rescinded. Occasionally, a ludicrous rule would surface, such as a permit "not to discharge" water into a stream; this one was renewable! Among Colorado mining men it was said that the secret to beating such regulations was to write letters to the government; the least that might be accomplished was to buy a year while the complaint was studied.

To make matters worse, hot fights developed at the state level over strip-mining regulations and Governor Richard Lamm's proposed severance tax on Colorado mineral production. From small producers and large, cries of protest reached the legislature during the 1975 and 1976 sessions. After much legislative indecision, the severance tax question was placed by initiative on the November 1976 ballot for the voters to decide. It seemed that just when mining was making a strong comeback, forces were rising to curb it.

Certain segments of Colorado mining were making solid headway, the best in decades. Gold prices mounted to nearly $200 an ounce, then dropped back, while silver reached the $4 to $5 level, both previously unheard of figures. Also, as of December 31, 1974, for

TABLE 4
1858 Through 1974
Cumulative Production

Gold	$ 944,911,901
Silver	655,052,350
Molybdenum	1,757,509,880
Uranium	1,008,194,946
Tungsten	100,567,591
Vanadium	261,717,841
Zinc	561,124,057
Coal	2,521,838,746
Oil	3,314,885,091
Oil Shale	2,091,349

Source: Colorado Division of Mines

195. Fieldwork and exploration are still a part of mining. These geologists are looking at a molybdenum prospect. Unlike the old days, they probably rode to the site in a four-wheel-drive vehicle. Helicopters have also been used to transport both men and equipment to inaccessible locations. Courtesy American Metal Climax, Inc., Climax.

196. Dynamite sticks were tapped in carefully. Now they are detonated electrically. Better ventilation makes it possible to work a mine sooner after blasting, but still one shift will drill and blast, and the next will muck. The men are working in Standard Metals' operation at Silverton. Courtesy Mike McRae, Durango.

197. Whether it is 1868, 1912, or 1976, the mining accident produces an agony of waiting for those on the surface. This group awaits news of miners trapped in a La Plata County coal mine. Courtesy the *Durango Herald*, Durango.

the first time since 1934 Americans could legally buy, sell, and hold gold. Consequently, production of both minerals climbed steadily. Silver production rose from $5 million in 1970 to $11.5 million in 1974, and gold from $1.3 to 7.6 million during the same period. The state's leading gold and silver district was the San Juans, with the Standard Metals operation near Silverton number one.

Dry statistics do not make interesting reading, unless they happen to be production figures for a mine in which the reader is financially interested; they do, however, indicate the trends of the seventies. Vanadium and uranium continued their decline, both suffering more than a fifty-percent reduction until 1974 when both rebounded strongly. Molybdenum wobbled, slipping slightly by mid-decade, although still maintaining its front rank. Zinc moved upward, but copper and tungsten declined slightly. Altogether, in 1974, metals accounted for thirty-two percent of Colorado production, with molybdenum, zinc, uranium, vanadium, and silver the top five.

The startling increases came not in the metals area but in the mineral fuels, which reflected the concern for national self-sufficiency. Production of crude oil multiplied over three times, gas nearly four, and coal by half again compared to 1969. Mineral fuels produced fifty-five percent of the total, with petroleum far in the lead. Thanks to oil, Colorado's total mineral value output increased for the tenth consecutive year, setting an all-time record of $697.9 million. The remaining production came from the nonmetallics; sand and gravel headed that list, followed by cement and limestone.

Coal mining rebounded, with strip mining receiving the most attention in the press and public forum. Visions of Appalachia strip mines haunted the industry—a past it could not live down. Such widely separated towns as Craig, Durango, and Oak Creek were being surveyed for coal deposits, both the underground and stripping varieties. Craig had two power plants under construction, ready markets for its coal. Already dragline buckets taking hundred-ton bites were stripping coal for customers as far away as Illinois. There was even speculation that a railroad would be built to Durango from the south, a long-ago dream, to transport coal to market. Such developments, planned or started, were bound to have serious consequences for the regions under discussion. Much soul-searching had already taken place to forestall them and to provide safeguards.

198. For the jeeper and tourist, mining has left a heritage to examine and, regrettably, to cart off. The camps, the mines, the cabins are disappearing as an increasing flood of visitors washes over them. On a quiet summer day in 1974, Alta gives a fleeting glimpse of what transpired during its mining days. Courtesy Richard Gilbert, Durango.

Labor disputes, both local and national, still dogged the coal industry. In late 1974 the United Mine Workers struck over national issues and contracts, pulling their Colorado locals with them. Throughout the strike some unaffiliated mines continued to operate, but most closed, showing clearly that the miners as well as the owners were caught up in broader issues. Eventually negotiations proved successful, and the miners returned to work. Fortunately, the strike did not last long enough to disrupt seriously the miners' lives or the economy. Strikes have become a way of life in the coal mines; as one person said, "We have one every three years, plus a few wildcat strikes in between for various grievances."

As mid-decade was reached, there were rumors that the Cripple Creek, Central City, and Georgetown areas might be reopened for serious mining efforts. Talk was cheap, action dear. The cost of opening and operating a mine continued to mount, keeping it primarily the game of the well-financed operator or the large company. The state was still short of smelting facilities, a decades' old complaint. Norman Blake, deputy commissioner of mines, wrote in 1973: "There are several hundred small mines in Colorado which could be opened up with very little capital investment . . . if we had the sampling works and mills." He went on to bemoan the fact that "our closest smelters are hundreds of miles away," too far to ship crude ore. He only echoed sentiments heard on so many previous occasions.

The giant on the Colorado mining scene in 1976, as the state celebrated its one hundredth birthday, was the Climax Molybdenum Company, itself a division of American Metal Climax. It opened the Henderson Mine after an investment of $500 million and fifteen years of exploration, planning, and construction. Hop-

TABLE 5
1975 Production

Molybdenum	$146,636,557
Silver	12,512,045
Gold	10,112,265
Coal	87,909,066
Crude Oil	365,632,619
Oil Shale	1,414,662

Source: Colorado Division of Mines

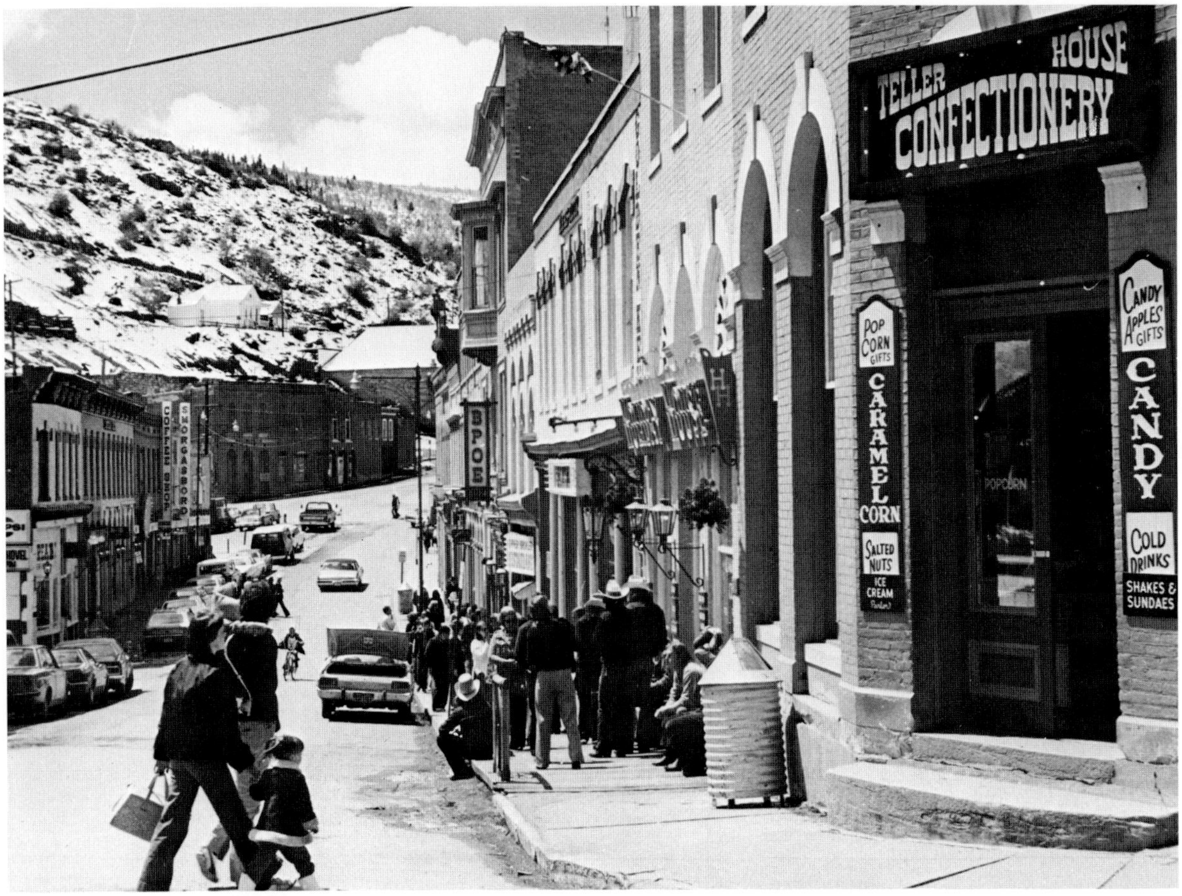

199. Central City and a few others have been converted into communities that mine tourists. Main Street has changed since the 1860s, but not much since the fire of 1874. Tourist attractions, restaurants, and souvenir shops have replaced the earlier merchants' stores. Courtesy Richard Gilbert, Durango.

ing to leave behind a better heritage, the company worked with environmentalists to plan reclamation *before* the damage had been done. Meanwhile, it carried on a continual search for new deposits, that included a drilling rig in the Weminuche wilderness area of the San Juan mountains. Under close environmental controls, AMAX and other companies also faced a government-imposed cut-off date of 1983 for mineral exploration in wilderness areas. So much had mining changed by Colorado's centennial.

Despite the fight over environmental controls, severance taxes, and government regulations, Colorado mining was on a firmer foundation than it had been since the nineteenth century. Whereas gold and silver had reigned over mining almost exclusively a century before, diversity now provided the key to success. The small operator was a vanishing breed, threatening to go the way of the prospector and his mule, but he was not finished and might yet revive. Women engineers and miners entered the mines, something which would have dumbfounded their male predecessors, who superstitiously considered their presence a bad omen.

Mining had changed, but much of it endured, only slightly updated. As the writer of "Ecclesiastes" explained: "What has been is what will be, and what has been done is what will be done; and there is nothing new under the sun." Men still worked far underground with machines, drills, shovels, and cars. Dust, danger, and fatigue continued to be their daily companions as they mined ore in a world nearly isolated from what was taking place on the earth above them. Where once alcohol had been a problem ("miners, by their nature, always have been pretty heavy drinkers," commented an official of the Colorado Division of Mines), marijuana now almost equaled it as a safety hazard in the mines. Reflecting the rest of society, studies showed that drug abuse more generally affected the younger miners and alcohol the over-thirty-five crowd. Miners and their families lived in Colorado mining towns filled with historical fact and fiction —Leadville, Silverton, and Red Cliff—much as they had done for generations, or in towns like Craig and Montrose, which had seldom before seen a resident miner.

In the 1970s Americans had a penchant for bumper stickers. One that appeared occasionally spoke volumes about what had transpired and what was yet to come: "Mining Is Everyone's Future."

200. Not all that remains of the mining era has been abandoned or ignored. The Maxwell House, in Georgetown, cited as one of the outstanding examples of Victorian residential architecture in the United States, is also one of the best preserved. Courtesy C. Eric Stoehr, Dallas.

201. Lonely monument to a vanished era—Idaho Springs' Argo mill. Scrap drives, abandonment, and vandalism have taken their toll. Vandalism is nothing new; a man complained in 1868 that he had to store everything that was loose around his inactive mill to keep it from being stolen. Courtesy C. Eric Stoehr, Dallas.

202. America's and Colorado's future fuel needs ride strongly on the outcome of exploration for oil and the reworking of older fields. This pump is working at the McCallum field on the windswept Colorado Plateau. Courtesy the Continental Oil Company, Denver.

Epilogue

"Mining Is Everyone's Future." Undoubtedly, some Coloradans would quibble with that statement, but it is valid historically. Today it is especially true, as the energy crunch emphasized, and tomorrow it will remain significant. Colorado mining affects all Coloradans, from the urban Denverite to the Telluride miner, whether or not they choose to recognize it.

That mining has played a significant role in Colorado's development since 1858 is an axiomatic statement, one which needs no elaboration at this point. The reader should be well aware of mining's importance in creating the territory, promoting settlement, providing a statewide transportation system, attracting investment, supporting cultural transmission, and encouraging a diversified economy. Aside from the money pumped directly into Colorado's economy through jobs, materials purchases, and reinvestment of mining profits, all these benefits accrued. These things were not accomplished without some deleterious results. All those people could not settle in a fragile ecological system or develop an exploitative industry without some damage to the environment. Nor could management and the workers hope to move together, when they marched to a different beat; when corporation mining emerged, so did union organization. There are both positive and negative examples in the history of mining in Colorado, and they need to be weighed against each other to arrive at conclusions about its total impact and significance.

Mining historian Rodman Paul called Colorado "the wonder of the mining West," a sentiment with which few miners would disagree. In the United States, Colorado is among the top three producers of molybdenum, vanadium, gold, silver, and tungsten, and is not far behind in lead, uranium, and zinc. Add to this coal, oil, natural gas, copper, sand, and gravel, and the breadth of this mineral treasure chest is apparent. The state has had an unusually long history of profitable mining, in an industry that cannot replace its product. Of all the western mining states Colorado stands alone in these respects.

The exploitation of major portions of these deposits has been achieved largely because of technical and scientific advances in mining and milling. Colorado has stood at the forefront of research, invention, development, and acceptance of improved methods and equipment. Nevada's Comstock was once the school for miners and engineers; Colorado has continued to fill that role.

The mining camps and towns, which have attracted more antiquarian, scholarly, and tourist interest than mining itself, have likewise made important contributions. It may be argued that most have acquired a ghost town epitaph, or exist as mere shadows of their once bustling selves. Can these remains be called only human and material wastes, blights upon nature's beauty? Perhaps, but to stop with this definition would be patently unfair. These communities and their people made significant contributions. The cultural transmission spoken of earlier came primarily under their auspices; for the immigrant, they constituted his educational experience and were melting pots of Americanism. Those earlier rushers were not refugees from advancing urbanism, they were its exponents. They prompted an ever widening economic system (agriculture comes quickly to mind), which remains. The decline and disappearance of places where once hopes and dreams could be realized, where the routine of daily living prevailed, and where a new generation was born, provided Colorado with tourist attractions. These things of themselves balance the waste that came with settlement.

A direct spin-off of mining was the establishment of satellite communities for supply, milling, or transportation: Denver, Boulder, Gunnison, and Durango are examples. Situated in key geographical locations, they have emerged as urban centers and have played roles in Colorado far beyond those originally intended. Also related directly to mining was Colorado's nineteenth-century growth. The federal census takers counted 34,277 people in 1860, largely the residue of the 1859 and early 1860 rushes. The troubles of the sixties were evidenced by a population increase of only 5,000. But then in the silver seventies, the number jumped to 194,000, and it reached 539,000 by 1900. Neighboring Wyoming, with little mining and much ranching, climbed from 9,000 to 92,000 during the same years.

Mining and mining camps exemplified much of the best of Colorado's diverse historical periods since 1858. Much of the optimism, initiative, and determination which characterized the nineteenth century came from mining, as did adaptiveness and ability to weather hard times during the twentieth century.

On the other side of the coin were speculation, impatience, tastelessness, and lack of foresight in Colorado mining communities and in mining. The industry passed through a number of cycles, the downswings hurting the state economically, politically, and socially. For the Indians who lived here, mining ended a traditional way of life in a series of episodes that cover neither of the conflicting sides with glory. To the on-rushing miners the end justified the means. Today it is easy to criticize their callousness, but it is hard to be sure one would have acted any differently at that time.

Colorado mining *was* economically motivated. Even in the camps the manifestations of civilized society can be traced partially to economic business factors. Many refinements were imitative of the East rather than innovative, as has been claimed. There was rarely time to reflect, innovate, and weigh long-range impact in the mining West. The result often was a veneer, rather than substance.

The miners of the nineteenth century had a better sense of history than some of today's mining critics. They realized that they were building for a better tomorrow, and, within the era's limitations, attained what they set out to accomplish. At first, perhaps, few expected to stay more than a season or two, but the idea of permanent settlement soon took hold. Those early Coloradans looked upon mining as a democratic opportunity, open to all who were willing to work; it soon became big business, and then Colorado expanded at unexpected rates. The prospectors and miners found that mining engendered expenses that were often uncompensated by the amount of ore found. Much of the real profits disappeared into the hands of merchants, railroad companies, freighters, and other businessmen who in turn invested in Colorado.

There has never been anything in American experience quite like the gold rushes of 1849 and 1859, nor any western immigrant quite like the mineral seeker. The motivation of these men might have been similar, but their pace, numbers, and impact were not. Historian Earl Pomeroy, writing about Pacific slope mining, had this to say: "Whatever the miners had been at home, the mining country as a whole included samples of all varieties of human nature as well as of all Satan's principal reagents for human metallurgy." He might have added that there was something in the excitement, gamble, and adventure of mining that took hold of people and would not let them go, thus shaping generations since 1858–59.

What the gold rush started, mining has continued. The development of Colorado dates from its beginnings, and mining has, to a great degree, submerged itself in the total Colorado economy and history. It has naturally played a less important role in the twentieth century than it did in the nineteenth. But the future remains exciting, much as it did in 1859. In the early 1960s, A. H. Koschmann wrote in "The Historical Pattern of Mineral Exploitation in Colorado," that the future of mining depended on success in finding new ore bodies and in improving mining and recovery methods to permit economic exploitation of low-grade deposits; success in research to develop new uses for and greater utilization of mineral raw materials; and success in research to adapt more minerals to man's use. These have been the crucial success factors throughout Colorado mining, and the future depends on them. With the men who mine, they will determine what happens in the years to come.

An old miner-prospector I once interviewed summarized his mining experience this way: "It's a long way to hell." He obviously dwelt upon the work, worry, and tentative rewards. There is more to it than that. There is a certain feeling, a spirit that is hard to capture. A young woman, Dell Merry, living in the declining camp of Caribou in the 1880s, came as close to it as perhaps one can with mere words. She wrote, "We miners are always looking for something big whether we get it or not. No risk, no gain is our motto. We have enough to live on and that is more than a good many have, so we ought to be thankful."

At almost the same time Dell was writing, the better known Horace Tabor gave this forecast for Colorado mining:

> In my opinion the mining industry will continue permanent in Colorado. There are going to be many valuable mines found yet in Colorado that are sleeping today. . . .The production of Colorado, more than that of any other country, will increase rather than diminish.

They were all accurate in their way—the miner, the housewife, and the owner. Colorado was founded on the bedrock belief that the mountains were so rich that even the fabled "King Solomon's mines" were not comparable. Glittering above everything was the dream of getting rich without work, a dream that proved illusionary. Still, Colorado did indeed assay out to be a mineral bonanza—a wondrous land.

Bibliographical Essay

It is hoped that the foregoing story has encouraged the reader to pursue Colorado mining further. The field is large, the challenge great, and the rewards satisfying if not financially enriching. A warning, however, is necessary: it is easy to get "hooked," and then the search for new and fascinating topics becomes endless. Colorado mining history has been prospected but not thoroughly mined, particularly for the twentieth century. Legends and romance are fine in their place, but it is time to look elsewhere for something of more significance. This essay does not attempt to list all the books, pamphlets, and articles that have been published since 1859. It is merely a personal narrative that offers some guidelines and comments which lead to other sources and open the doors to the excitement of Colorado mining.

No general history has been published. Rodman Paul's *Mining Frontiers of the Far West 1848–1880* (New York: Holt, Rinehart & Winston, 1963; reprinted by the University of New Mexico Press, 1974) and William Greever's *The Bonanza West* (Norman: University of Oklahoma Press, 1963) present the nineteenth-century story within the framework of the whole mining frontier. *The Colorado Magazine*, published by the State Historical Society of Colorado, has scores of articles on every phase of mining, from personal reminiscences to scholarly interpretations; most are concerned with the nineteenth century. A decided help is the index which carries through 1960. Among the older histories, Hubert Howe Bancroft's *History of Nevada, Colorado and Wyoming 1540–1888* (San Francisco: History Company, 1890), Frank Hall's *History of the State of Colorado*, 4 vols. (Chicago: Blakely Printing Co., 1890), Wilbur Stone's *History of Colorado*, 4 vols. (Chicago: S. J. Clarke, 1918), and Jerome Smiley's *Semi-Centennial History of Colorado*, 2 vols., (Chicago: Lewis Publishing Co., 1913) provide framework and some first-hand accounts. LeRoy R. Hafen, ed., *Colorado and Its People*, 4 vols. (New York: Lewis Historical Publishing Co., 1948) and James Baker, ed., *History of Colorado*, 5 vols. (Denver: Linderman, 1927) offer chapters on individual topics, including coal, oil, labor, and mining in general. For statistics and a county-by-county examination, Charles Henderson's *Mining in Colorado* (Washington: Government Printing Office, 1926) is invaluable through the early 1920s. The question of total production and value will never be resolved, but Henderson gives a very scholarly estimate for the period up to the publication of the book.

On more specific topics within mining, the following are noteworthy. Clark Spence offers interesting insights in his *Mining Engineers & the American West* (New Haven: Yale University Press, 1970) and *British Investment and the American Mining Frontier* (Ithaca:

Cornell University Press, 1958). W. Turrentine Jackson has done the same in *The Enterprising Scot* (Edinburgh: Edinburgh University Press, 1968). Otis Young deals with the techniques and machines in his *Western Mining* (Norman: University of Oklahoma Press, 1970) and *Black Powder and Hand Steel* (Norman: University of Oklahoma Press, 1975) and Rodman Paul looks at Colorado in "Colorado as a Pioneer of Science in the Mining West," published in *The Mississippi Valley Historical Review* 47:34–50. See also A. H. Koschmann's "The Historical Pattern of Mineral Exploitation in Colorado," *Quarterly of the Colorado School of Mines* 57:7–25. Arthur Todd, *The Cornish Miner in America* (Truro: D. Bradford, 1967), and A. L. Rowse, *The Cousin Jacks* (New York: Charles Scribner's Sons, 1969) examine these first-rate miners. For those interested in mining unions, Richard Lingenfelter gives the background into the 1890s in *The Hardrock Miners* (Berkeley: University of California Press, 1974) and Vernon Jensen takes the story into the present century in his *Heritage of Conflict* (Ithaca: Cornell University Press, 1950). James Allen, in *The Company Town in the American West* (Norman: University of Oklahoma Press, 1966) gives some Colorado examples, and my own *Rocky Mountain Mining Camps* (Lincoln: University of Nebraska Press, 1974 reprint) provides general background.

Numerous pictorial studies have focused specifically on Colorado's mining camps and frontier. One of the most delightful is Muriel Wolle's *Stampede to Timberline* (Chicago: Swallow Press, 1974) which has a personal touch that is captivating. Also worthwhile is Don and Jean Griswold's *Colorado's Century of Cities* (Denver: Smith-Brooks, 1958). Robert Brown in his several volumes, which include *An Empire of Silver* (Caldwell: Caxton, 1965) has performed a valuable service in preserving the vanishing scene with his photographs.

For the beginnings of the Colorado gold fever, LeRoy R. Hafen's four edited volumes are recommended: *Pike's Peak Gold Rush Guidebooks of 1859* (Glendale: Arthur Clark, 1941), *Overland Routes to the Gold Fields* (Glendale: Arthur Clark, 1942), *Colorado Gold Rush Contemporary Letters and Reports 1858–1859* (Glendale: Arthur Clark, 1941), and *Reports from Colorado 1859–1865* (Glendale: Arthur Clark, 1961). Hafen's edited version of Henry Villard's *The Past and Present of the Pike's Peak Gold Regions* (Princeton: Princeton University Press, 1932) furnishes its readers with an excellent description of Colorado mining in 1859. For the 1860s, two contemporaries have left fine accounts: Ovando Hollister, *The Mines of Colorado* (Springfield: Samuel Bowles & Co., 1867) and Frank Fossett, *Colorado* (New York: C. G. Crawford, 1876, 1878, 1879, 1880). Fossett's work, issued in several editions, takes the reader through the 1870s. Both Hollister and Fossett have been recently reprinted (Hollister, New York: Arno Press, 1973; Fossett, New York: Arno Press, 1973 and Glorieta, N.M.: Rio Grande Press, 1976). Thomas Marshall has edited the *Early Records of Gilpin County* (Boulder: University of Colorado, 1920), a book that gives a good insight into mining districts and laws. Nolie Mumey has edited two volumes on the mining laws and proceedings of the Buckskin Joe district: *History and Proceedings of Buckskin Joe* (Boulder: Johnson Publishing, 1961) and *Early Mining Laws of Buckskin Joe, 1859* (Boulder: Johnson Publishing, 1961). In *Echoes from Arcadia* (Denver: Lanning Bros., 1903), Frank C. Young recaptures Central City in the 1860s and 1870s. Samuel Cushman and J. P. Waterman supplement Young with their *The Gold Mines of Gilpin County* (Central City: Register, 1876), a description of mining and milling during the same period. Cushman provides the same service for those interested in the Georgetown region with *The Mines of Clear Creek County* (Denver: Times, 1876).

Don and Jean Griswold have looked into the big excitement of the 1870s in *The Carbonate Camp Called Leadville* (Denver: University of Denver, 1951), as does the *History of the Arkansas Valley, Colorado* (Chicago: Baskin & Co., 1881). Newspaper editor Carlyle Davis reminisces in *Olden Times in Colorado* (Los Angeles: Phillips, 1916) and some of his statements need verifying, a warning applicable to all books of this type. This writer scrutinizes a lesser camp in *Silver Saga: the Story of Caribou, Colorado* (Boulder: Pruett, 1974) and John Horner peeks at Georgetown in *Silver Town* (Caldwell: Caxton, 1950). Contemporary directories provide all types of information. See Thomas B. Corbett, comp., *The Colorado Directory of Mines* (Denver: Rocky Mountain News, 1879) and Robert Corregan and David Lingane, comps., *Colorado Mining Directory* (Denver: Colorado Mining Directory Co., 1883). Mary Hallock Foote was in Leadville; for this educated woman's view, see Rodman Paul, ed., *A Victorian Gentlewoman in the Far West* (San Marino: Huntington Library, 1972).

The rest of the century has attracted its share of attention as well. Marshall Sprague's *Money Mountain* (Boston: Little, Brown and Company, 1953) traces Cripple Creek's history, as does Robert Taylor's *Cripple Creek* (Bloomington: Indiana University Press, 1966), though from a different viewpoint. A delightful way to capture the spirit of this period is to read Mabel B. Lee's *Cripple Creek Days* (New York: Doubleday & Co., 1958). Harriet Backus's *Tomboy Bride* (Boulder: Pruett, 1969) is equally absorbing for the Telluride area a decade later. Nolie Mumey, *Creede* (Denver: Artcraft Press, 1949) looks at one of the overrated camps, and Aspen is still seeking a historian.

Only a few of the mining men of the nineteenth century have received full-length biographical attention. Frank Waters's *Midas of the Rockies* (Chicago: Swallow Press, 1972) spotlights Cripple Creek's Winfield Scott Stratton. Horace Tabor, possibly the most notable personality of his generation, has attracted more attention, the most recent being this author's *Horace Tabor: His Life and the Legend* (Boulder: Colorado Associated University Press, 1973). Many of the multivolume Colorado histories mentioned earlier have biographical sections that give short accounts of various mining figures, as do some other volumes such as *History of Clear Creek & Boulder Valleys* (Chicago: Baskin & Co., 1880).

Turn-of-the-century Colorado mining has been drawing interest basically because of the study of labor unrest. George Suggs, *Colorado's War on Militant Unionism* (Detroit: Wayne State University Press, 1972), George McGovern and Leonard Guttridge, *The Great Coalfield War* (Boston: Houghton Mifflin, 1972), and Barron Beshoar's *Out of the Depths* (Denver: Golden Bell Press, 1957) have discussions of the hardrock and coal troubles. For a mining engineer's account of more peaceful days, check T. A. Rickard, *Across the San Juan Mountains* (New York: Engineering and Mining Journal, 1903). Gene Gressley, ed., *Bostonians and Bullion* (Lincoln: University of Nebraska Press, 1968) takes the reader to Telluride and its mines. After these, production is skimpy. Duane Vandenbusche and Rex Myers examine *Marble* (Denver: Golden Bell Press, 1970) and its unusual industry; Otis King gives his side of the Climax molybdenum story in *Gray Gold* (Denver: Big Mountain Press, 1959); and Kathleen Bruyn looks at the *Uranium Country* (Boulder: University of Colorado Press, 1955). Erl Ellis takes a photographic look at dredging in *The Gold Dredging Boats Around Breckenridge, Colorado* (Boulder: Johnson Publishing, 1967). There are other studies of this nature, but they are few and far between.

Don't despair if a favorite camp or district seems to have been overlooked. Go back to *The Colorado Magazine* in which almost all of them are mentioned, and where there are reviews of many books and pamphlets. The various editions of the Denver Westerners' *Brand Book* contain information. From this point, the research becomes more difficult and, consequently, more fascinating. Journal and magazine articles can be particularly informative, if up-to-date. Three invaluable books for the scholar are Olive Jones's *Bibliography of Colorado Geology and Mining* (Denver: Smith-Brooks, 1914), Virginia Wilcox's *Colorado, A Selected Bibliography of Its Literature 1858–1952* (Denver: Sage Books, 1954), and Richard Holt's *Bibliography, Coal Resources in Colorado* (Denver: Colorado Geological Survey, 1972).

Both *The Engineering and Mining Journal* (New York) and the *New York Times* have indexes, the former a great source for Colorado mining. *The Mining and Scientific Press* (San Francisco) is not as helpful for this area. The *Mining Record* (New York) is interesting for the 1880s but hard to find. Within Colorado, numerous mining publications have appeared, such as Georgetown's *The Mining Review* and Denver's current *The Mining Record*. All contain an abundance of information and should not be overlooked by the thorough scholar.

Newspapers are a mandatory source that must never be slighted, since the gold of mining research is to be found among their pages. Mining camp and town papers provide contemporary insights into every phase of life. A gentle reminder—beware of editorial bias. Donald Oehlert's *Guide to Colorado Newspapers 1859–1863* (Denver: Bibliographical Center, 1964) tells where many newspapers can be found but does not completely cover the possibilities.

Government publications, federal and state, began appearing well back in the nineteenth century, in both article and book form. The list is far too long to include here, but some outstanding examples follow. Rossiter Raymond's volumes on the *Statistics of Mines and Mining* (Washington: Government Printing Office, 1868–75), Samuel Emmons's *Geology and Mining Industry of Leadville, Colorado* (Washington: Government Printing Office, 1886), the *Minerals Yearbook* (earlier called *Mineral Resources*), and a myriad of Geological Survey papers reflect federal government involvement. The census records, both original and published, contain a wealth of information. Colorado, for example, has produced coal mine inspector reports, annual reports (now called summaries) of mining, and yearbooks. The state has also had its own geological survey, which published bulletins such as P. G. Worcester's *Molybdenum Deposits of Colorado* (Denver: Eames Bros., 1919) and R. George, *Oil Shales* (Denver: Eames Brothers, 1921). Occasionally one runs across county or town publications, but these tend to be promotional.

Other worthwhile sources to investigate are the libraries of the various Colorado schools which offer graduate degrees. For years graduate students have been probing Colorado subjects in theses and dissertations with varying results and readability, but this should not deter the determined researcher. A good place to start would be the University of Colorado, simply because it has been in the game longest.

To jump from government publications and graduate tomes, crammed as they are with facts and footnotes, to fiction might seem absurd, but very often a novel can give the spirit of the times better than dry facts, especially if the author has carefully researched his background material. Fiction came amazingly early, and a fine place to start is with Mary Hallock Foote's novels set in Leadville. She herself becomes the center of attention in Wallace Stegner's *Angle of Repose* (Greenwich: Fawcett Crest, 1972). David Lavender takes the reader into the San Juans in *Red Mountain* (New York: Popular Library, 1974) and there are others, ranging from abominable to nearly as good as those just mentioned.

After poring over fiction and nonfiction, secondary and primary published sources, the researcher's next step is to turn to the resources and holdings of the various research collections. The State Historical Society, the Western History Department of the Denver Public Library, and the Western Historical Collections of the University of Colorado are the big three. They contain diaries, letters, reports, interviews, photographs—the whole spectrum of records pertaining to Colorado mining. The Colorado State Archives have public records, and the Colorado Division of Mines holds reports on individual mines. There are other collections scattered throughout the state, including the Center of Southwest Studies at Fort Lewis College and the Arthur Lakes Library at the School of Mines. Many of the mining camps still in existence have their town records and perhaps other documents in the city hall, a rewarding place to research. Some businesses have saved their records, an example being the First National Bank of Denver, which has much to interest mining students. Many counties or towns have established historical societies, which may have material related to mining; a letter of inquiry could elicit rewarding information. Outside the state, the Henry E. Huntington Library and the Bancroft Library, both in California, the Newberry Library in Chicago, and the National Archives in Washington have important Colorado holdings.

If the foregoing tempts the reader to enter into the captivating field of mining history, then this volume has proven successful. The vista is limited only by the researcher's imagination. Like mining, the field has been thoroughly scratched and in a few places deeply penetrated; however, many nuggets still remain to be found, and the search for them provides its own rewards. Field investigation should not be overlooked, even if nothing of substance still stands, for one can gain a feel for the site and an understanding of geography that cannot be acquired by armchair investigation. Good luck and good researching!

Index

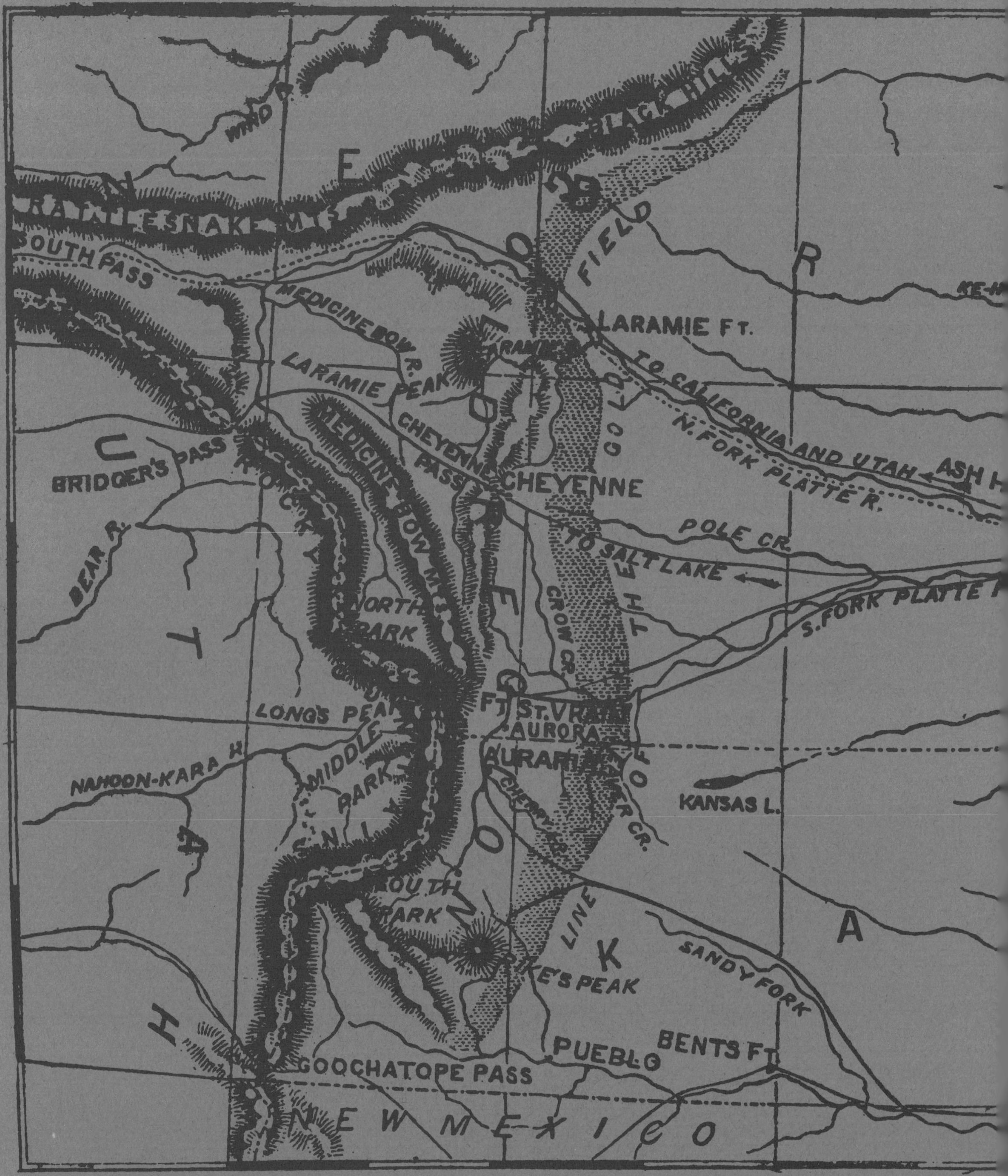
WIND R.
BLACK HILLS
RATTLESNAKE MTS
SOUTH PASS
MEDICINE BOW R.
LARAMIE FT.
LARAMIE PEAK
FIELD
TO CALIFORNIA AND UTAH
N. FORK PLATTE R.
CHEYENNE PASS
CHEYENNE
MEDICINE BOW MTS
BRIDGER'S PASS
BEAR R.
POLE CR.
TO SALT LAKE
S. FORK PLATTE
CROW CR.
NORTH PARK
LONG'S PEAK
AURORA
NAHOON-KARA R.
MIDDLE PARK
KANSAS L.
SOUTH PARK
PIKE'S PEAK
LINE
SANDY FORK
PUEBLO
BENTS FT.
GOOCHATOPE PASS
NEW MEXICO